# CONCRETE AND MASONRY

## REPAIRS AND UTILITIES

US ARMY

Fredonia Books
Amsterdam, The Netherlands

Concrete and Masonry Repairs and Utilities

by
United States Army

ISBN: 1-4101-0839-2

Copyright © 2005 by Fredonia Books

Reprinted from the government edition

Fredonia Books
Amsterdam, The Netherlands
http://www.fredoniabooks.com

# CONTENTS

# SECTION I

# INTRODUCTION

## 1. Purpose and Objectives

This manual is a materials and performance guide for use by installation personnel engaged in the construction, maintenance, and repair of concrete and masonry structures or elements of structures other than pavements, to assure:

*a.* Proper selection and employment of appropriate materials, equipment, and practices.

*b.* Competent inspection and appraisal of work done.

*c.* Elimination of hazards to life and property.

*d.* Preparation and utilization of records and reports.

## 2. General Application

The guidance in this manual comprises descriptions and discussions of materials, equipment, and methods used in construction, maintenance, and repair of concrete and masonry. It applies to housing, warehouses, technical buildings, special buildings, water towers, control towers, water supply, sewerage installations, and other structures. (For concrete roads, runways, and miscellaneous pavements, see TM 5–624.) Supplementary information useful to individuals concerned with concrete and masonry is contained in reference publications, especially other technical manuals of the TM 5–600-series, listed in appendix I. These reference publications should be readily available. Additional details on concrete, masonry, waterproofing, and dampproofing may be found in appendixes II through V.

## 3. Inspection

The production or repair of concrete and masonry work in such a fashion as to yield structures that will provide service for the anticipated period of use depends on the selection and employment of appropriate materials and practices. Satisfactory structures frequently are not obtained even though the materials provided and the practices selected are such as should yield satisfactory results. The reason for failure to achieve the desired results is usually attributable to inadequate supervision or inspection. The cost of competent personnel is relatively small compared with the resulting insurance of quality of the structure. Competent inspection prevents mistakes and permits economical use of materials. Clear-cut specifications are necessary to provide for an efficient and smooth-working job. To supplement the specifications, the foreman or inspector should be provided with a working library of appropriate reference publications including such of these listed in appendix I as are applicable to his responsibilities. The inspector should be informed of any special problems that are likely to be met on the job. The American Concrete Institute Manual of Concrete Inspection states (par. 2): "The small job needs inspection even more than the large job.' Inspection of concrete and masonry construction includes the following:

*a.* Identification, examination, and acceptance of materials.

*b.* Control of proportioning and batching of materials.

*c.* Performance of tests on samples.

*d.* Examination of foundation excavation, preparation of bases, forms, and other work.

*e.* Continual inspection of storing, mixing, conveying, placing, consolidation, finishing, and curing of materials.

*f.* Preparation of records and reports.

# SECTION II

# EXCAVATING, PREPARING BASES, FILLING, AND BACKFILLING

## 4. General

Figure 1 shows a suggested method for laying out a foundation preparatory to excavating.

## 5. Depths of Excavations

*a. General.* From the working drawings, which usually show existing grades, proposed finished grades, and elevations of first floors, footings, and foundations, determine the amount of excavating, filling, and backfilling needed. If the drawings do not show existing and finished grades, depths of footings may be determined from soil capacities below the frost line, or from the following guide:

| Temperature zone (F.) | Minimum depth below existing grade (feet) |
|---|---|
| +20° | 2 |
| 0 | 3 |
| −20° | 4 |

Take care not to excavate below prescribed depths, but if this happens, place concrete to the depth actually excavated. Do not refill excavations to the specified depth before placing concrete—it is too difficult to compact the fill surface properly. Since forms are not needed for footings in stable soils such as clay or disintegrated rock, footing excavations in these soils may be the actual width of footings as shown on drawings.

*b. Foundation Soil Tests.* Figure 2 shows a loading platform for foundation soil tests. Seldom, however, will the post engineer be required to make such tests.

*c. Foundation Design Loads.* Foundations should support about twice the design load without showing excessive settlements. Consult design handbooks and local city building codes for allowable design bearing values for various types of foundation materials for certain areas. Also, design bearing values for soil formations are generally available from District Engineers for installations within their areas. In the absence of data from these sources, values shown in figure 3 may be used.

KEY

A'

B' 8' A

C' 6' C D'

10'

E' B F'.

B'

8'-0"      6'-0"

10'-0"

TRENCH PARTLY EXCAVATED

PLUMB BOB

OUTSIDE LINE OF FOOTING
OUTSIDE LINE OF WALL
INSIDE LINE OF WALL
INSIDE LINE OF FOOTING

BATTER BOARDS

TOP OF BATTER BOARDS
TO BE LEVEL WITH TOP
OF FOUNDATION WALLS

INSIDE LINE OF FOOTING
INSIDE LINE OF WALL

OUTSIDE LINE OF WALL
OUTSIDE LINE OF FOOTING

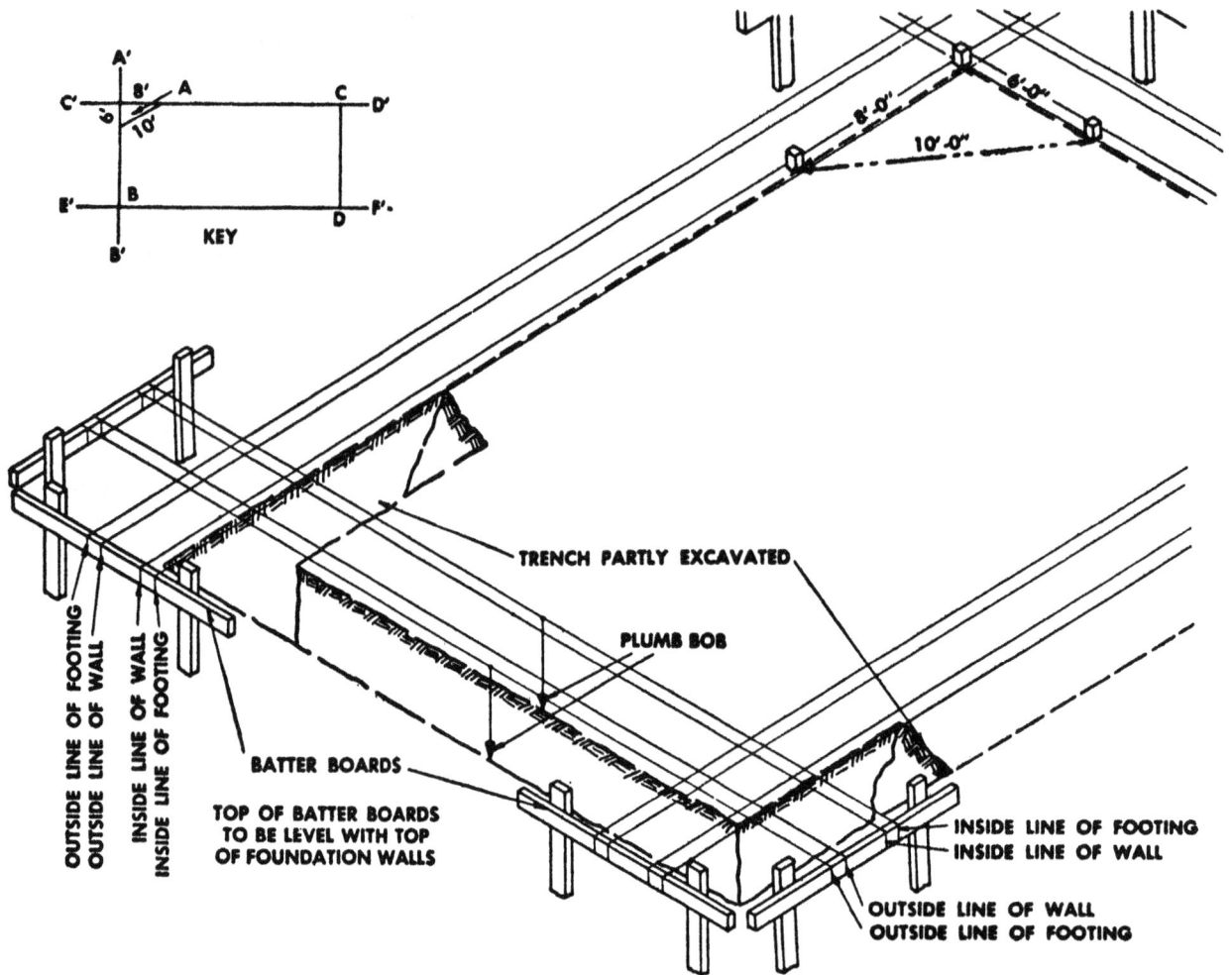

Step 1. Establish base line A'B' of indefinite length, locating face of front wall of building.
Step 2. Locate A and B, corners of building.
Step 3. Run lines C'D' and E'F' through A and B, at right angles to A'B' and of indefinite length. To form right angle, establish 10-foot diagonal between A'B' and C'D', 6 and 8 feet respectively from A as shown.
Step 4. Measure length of building from A and B to locate points C and D, then run line CD.

Step 5. As a check, measure AD and BC; they should be equal.
Step 6. Mark exact locations of A, B, C, and D on stakes driven in ground.
Step 7. Erect Batter boards as shown, leveling all top lines with line level placed on chalk line at midpoints of building corners. Locate markings and notches with chalk lines and plumb bob. Make all horizontal measurements perfectly level.

*Figure 1. Suggested method of staking out foundation.*

4

TO POST

7'-0"

12" | 2'-6" | 2'-6" | 12"

TWO ⅝" x 8" LAG SCREWS IN END EACH 6" x 8"

B

3" x 8" UNDER

2" x 12" PLANK FLOORING SPIKED TO TIMBERS WITH 60d COMMON NAILS

7'-0"

A ← → A

6" x 8" UNDER | 6" x 8" UNDER | 6" x 8" UNDER

6" x 8" UNDER | 6" x 8" UNDER | 6" x 8" UNDER

3" x 8" UNDER

B

TO POST

TWO STRANDS OF 6 GAGE GALVANIZED WIRE TWISTED FOR ADJUSTMENT

APPROX 18'-0"

**PLAN**

TOP OF PLATFORM TO BE APPROX 2'-0" ABOVE EXISTING GRADE AT START OF TEST

TWO 2" x 6" SCABS EIGHT 4½" LAG SCREWS

¼" x 2½" STRAP WITH FOUR ⅜" LAG SCREWS

⅜" DIA EYE BOLT

SEE DETAIL FOR WIRE GUARD

PIANO WIRE

GUARD

TWO 2" x 10"

SCALE

30°

6" x 6"

6" x 6"

1" DIA 3'-0" INTO GROUND, ⅛" SAW CUT TO HOLD WIRE

THREE ¾" DIA BOLTS

TWO ½" DIA BOLTS

PAIRS ½" DIA BOLTS AT 12" CC

5'-0" MIN

VARIES

VARIES — 4'-0" MIN OR ELEVATION OF BOTTOM OF FOOTINGS

5'-0" MIN

6" POST 5'-0" IN GROUND

12" x 12" x 1½" STEEL PLATE TRUE AND SQUARE

**IMPORTANT**

This platform with base plate weighs approximately 1000 pounds when built from timber that weighs about 40 pounds per cubic foot.

**SECTION A-A**

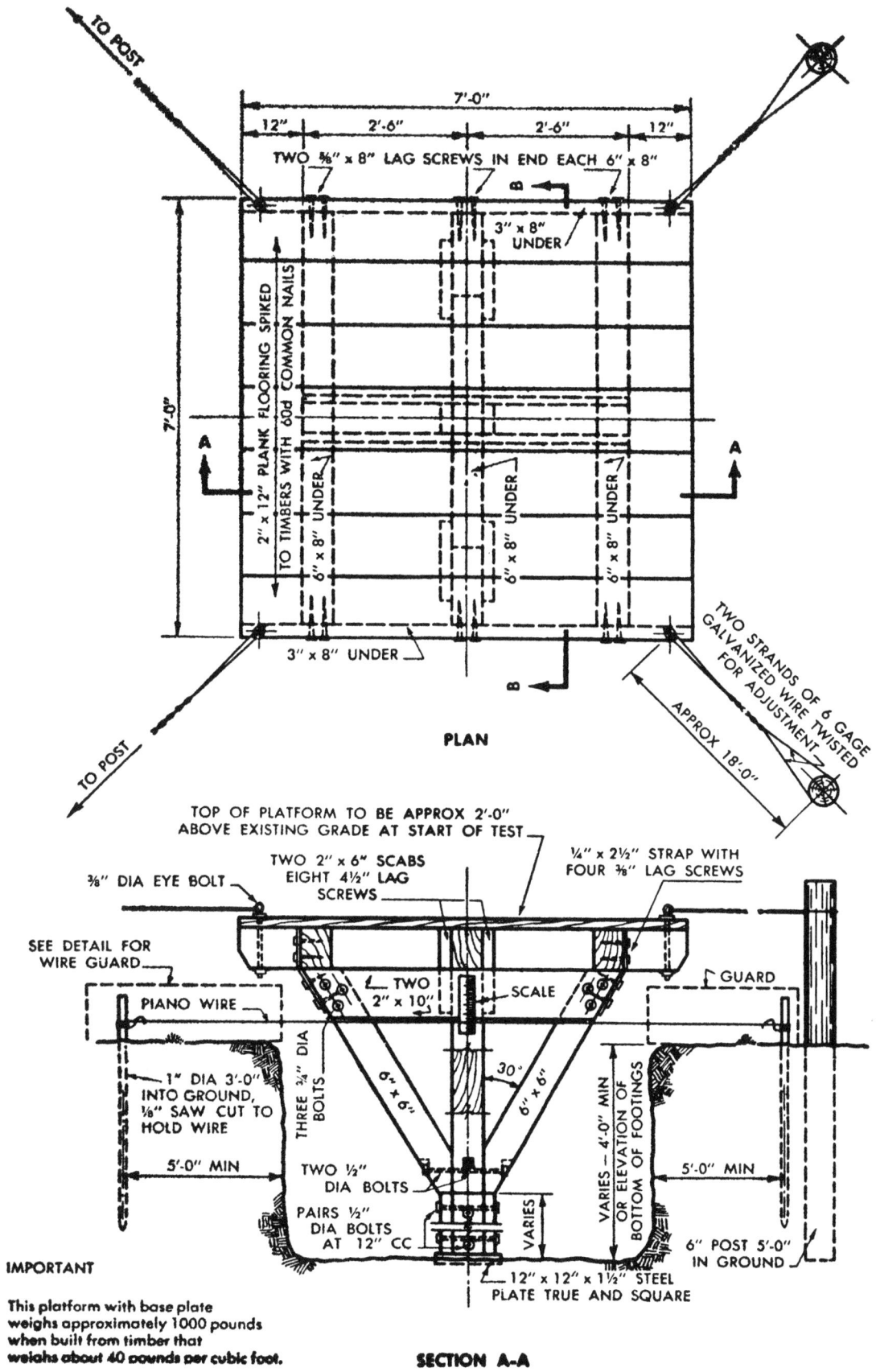

*Figure 2. Loading platform for foundation soil tests.*

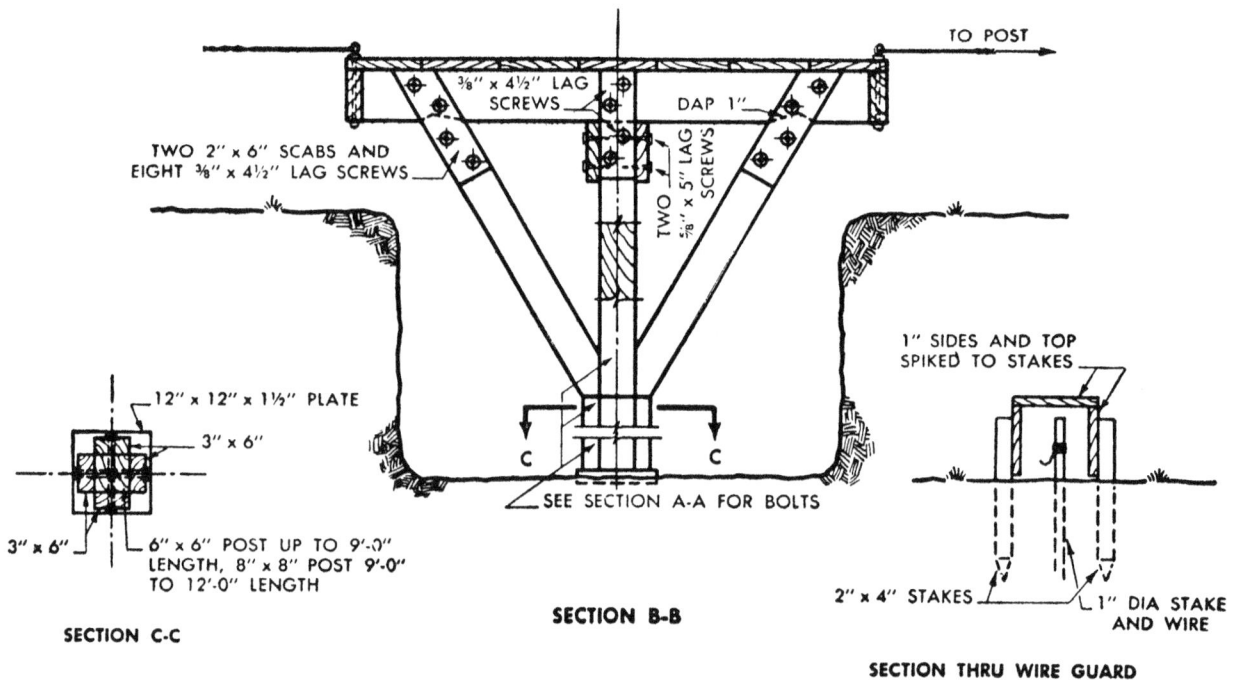

TO POST

³⁄₈" × 4½" LAG
SCREWS

DAP 1"

TWO 2" × 6" SCABS AND
EIGHT ³⁄₈" × 4½" LAG SCREWS

TWO ⅝" × 5" LAG SCREWS

12" × 12" × 1½" PLATE

3" × 6"

3" × 6"

6" × 6" POST UP TO 9'-0"
LENGTH, 8" × 8" POST 9'-0"
TO 12'-0" LENGTH

SEE SECTION A-A FOR BOLTS

1" SIDES AND TOP
SPIKED TO STAKES

2" × 4" STAKES

1" DIA STAKE
AND WIRE

**SECTION C-C**

**SECTION B-B**

**SECTION THRU WIRE GUARD**

NOTE

In construction of platform, take care
that all framing is accurately and neatly done,
and full bearing obtained between all parts.
Draw all bolts, lag screws, and so on, up tight.

PROCEDURE

In making soil test with this platform, the base plate must have full bearing and rest level on the soil to be tested, which must be undisturbed. When setting the platform on plate, care shall be taken to avoid jarring and workmen should keep out of the excavation as much as possible.

During the entire period of the test, surface and rain water must be kept out of the excavation by ditching, banking and tarpaulins.

Material used for loading should be carefully weighed and placed in uniform layers in such a manner as to eliminate as far as possible eccentricity of load at any time.

Platform should be loaded in two thousand pound increments at 24 hour intervals and the settlement carefully recorded immediately before and after each loading. Initial reading to be made immediately after platform is set up.

If 2" settlement is recorded before the capacity load of 10 tons is in place the test may be stopped. If not, the 10 ton load should be left in place for 10 days and daily settlement readings made. The iron stakes holding the indicator wire should be carefully tied into a bench mark and checked from time to time during the test.

In all cases the thickness of the stratum tested and the character of the underlying strata should be investigated, either by driving down a ⅝" rod, wash or core borings or open pit. In many cases the rod will give fairly accurate information by the speed and amount of penetration under repeated hammer blows. Borings will give more accurate information and should be used where the magnitude of the work requires or where doubt exists. The open pit, while more costly, allows visual examination of the material.

*Figure 2*—Continued.

| CLASS OF MATERIAL | MINIMUM DEPTH OF FOOTING BELOW ADJACENT VIRGIN GROUND | VALUE PERMISSIBLE IF FOOTING IS AT MINIMUM DEPTH. POUNDS PER SQUARE FOOT | INCREASE IN VALUE FOR EACH FOOT OF DEPTH THAT FOOTING IS BELOW MINIMUM DEPTH. POUNDS PER SQUARE FOOT | MAXIMUM VALUE. POUNDS PER SQUARE FOOT |
|---|---|---|---|---|
| 1 | 2 | 3 | 4 | 5 |
| Rock | 0' 0" | 20% of ultimate crushing strength | 0 | 20% of ultimate |
| Compact coarse sand | 1' 0" | 1500* | 300* | 8000 |
| Compact fine sand | 1' 0" | 1000* | 200* | 8000 |
| Loose sand | 2' 0" | 500* | 100* | 3000 |
| Hard clay or sandy clay | 1' 0" | 4000 | 800 | 8000 |
| Medium stiff clay or sandy clay | 1' 0" | 2000 | 200 | 6000 |
| Soft sandy clay or clay | 2' 0" | 1000 | 50 | 2000 |
| Adobe | 1' 6" | 1000** | 50 | |
| Compact inorganic sand and silt mixtures | 1' 0" | 1000 | 200 | 4000 |
| Loose inorganic sand and silt mixtures | 2' 0" | 500 | 100 | 1000 |
| Loose organic sand and silt mixtures and muck or bay mud | 0' 0" | 0 | 0 | 0 |

\* These values are for footings one foot in width and may be increased in direct proportion to the width of the footing to a maximum of three times the designated value.

\*\* For depths greater than eight feet (8') use values given for day of comparable consistency.

*Figure 3. Average allowable soil pressure (pounds per square foot).*

## 6. Drainage

Slope or ditch the ground surfaces around buildings and structures to keep water from running into excavated areas. Remove all water from the excavation before building forms and placing concrete.

## 7. Fencing and Shoring

If not already protected, erect substantial railings near the edges of excavations to safeguard personnel and property. Shore excavation side walls against lateral movement and cave-ins, if the type of soil so requires (fig. 4). Inspect buildings and structures adjacent to new excavations to determine the need for shoring or other protective measures.

## 8. Preparing Base for Slabs on Grade

a. Bases for concrete slabs on grade consist of well compacted fill of crushed stone, sand, gravel, or cinders which have been wetted down and tamped to designed grade and line. Cinders when wet produce acids which are destructive to some types of utility lines; take adequate precautions to prevent such damage.

b. The base includes forming for utility ducts or drainage trenches, interior integral footings, and equipment or machinery pads as needed. Sewage lines, drain lines, and other utilities may be placed in the fill. Fill should be placed on stable undisturbed bearing soil. Subsequent to grading and prior to concrete placement the fill shall be suitably treated to protect against insects and rodents (See TM 5–632).

EARTH LEVEL                    2" x 8" PLANKS

4" x 6" BRACES                    4" x 6" BRACES

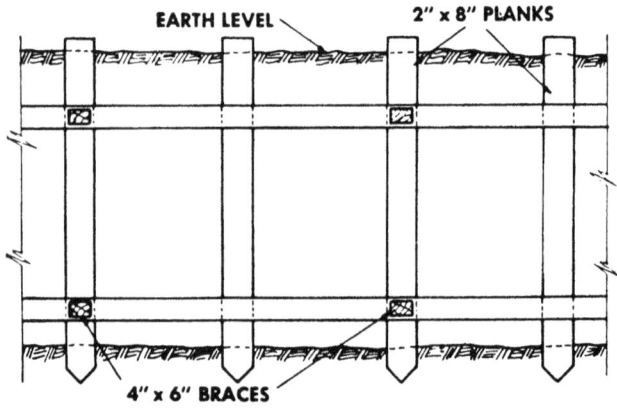

**SKELETON STAY BRACING**          **HORIZONTAL STAY BRACING**

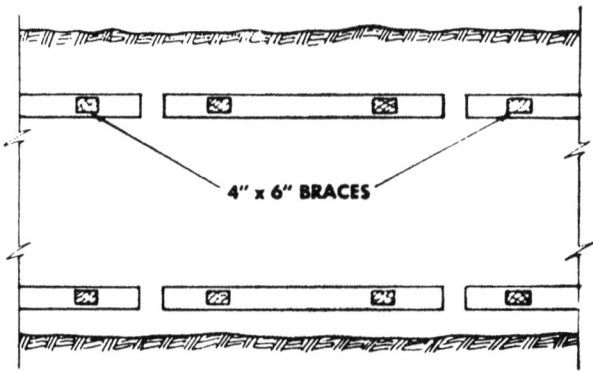

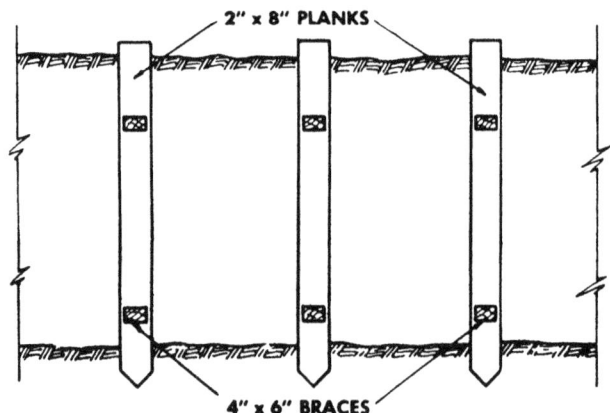

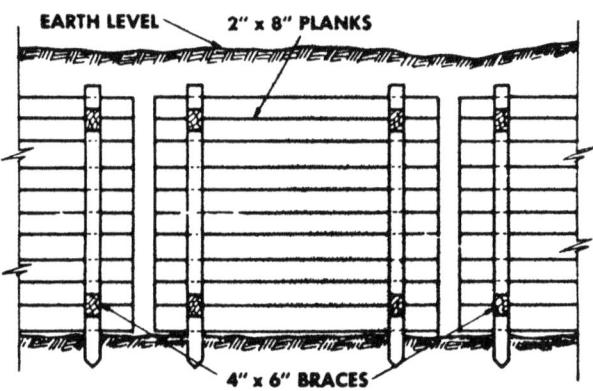

2" x 8" PLANKS                 EARTH LEVEL    2" x 8" PLANKS

4" x 6" BRACES                    4" x 6" BRACES

**VERTICAL STAY BRACING**          **HORIZONTAL SHEETING**

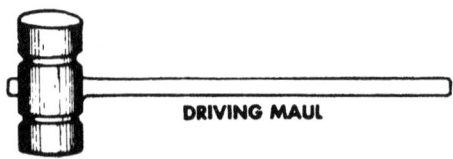

**DRIVING MAUL**                   THREADED

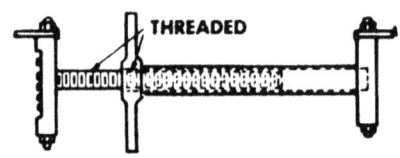

                                   **EXTENSIBLE TRENCH BRACE**

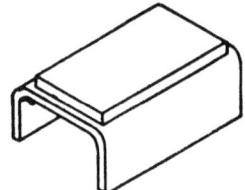

**STEEL DRIVING BLOCK**

1  Trench-bracing methods

*Figure 4.  Bracing and shoring.*

AGO 10179A

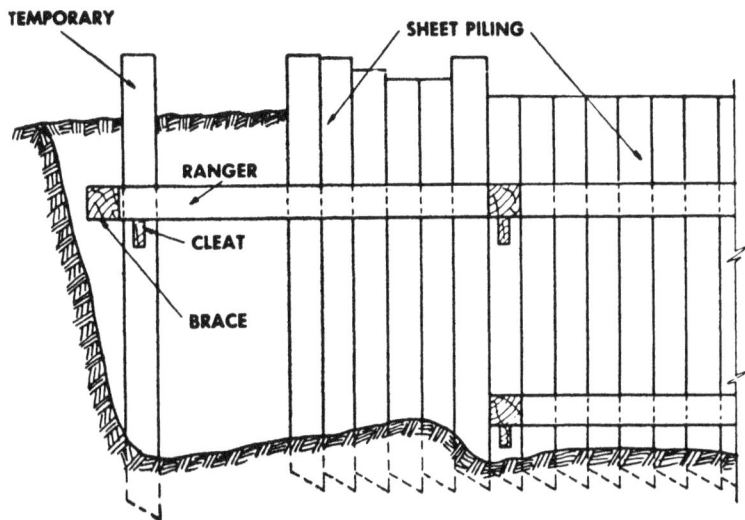

**SECTION OF A TRENCH SHEET-PILING METHOD**

TEMPORARY

SHEET PILING

RANGER

CLEAT

BRACE

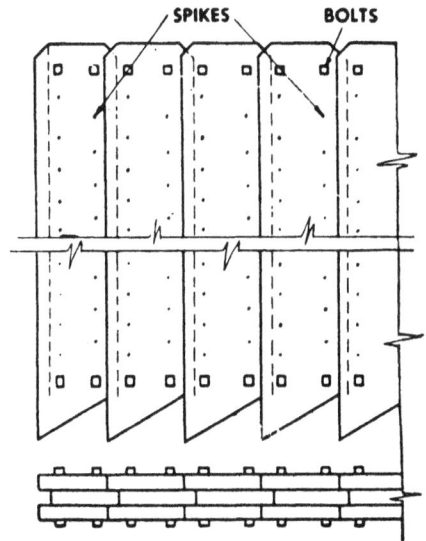

**WAKEFIELD SHEET-PILING WOOD**

SPIKES

BOLTS

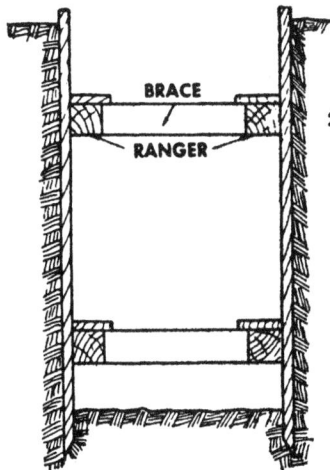

**TRENCH CROSS SECTION SHOWING SHEET PILING**

BRACE

RANGER

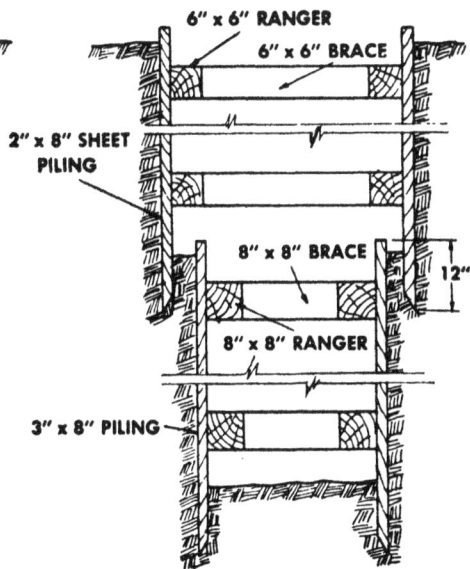

**TWO-SECTION SHEET PILING DEEP EXCAVATION**

6" x 6" RANGER

6" x 6" BRACE

2" x 8" SHEET PILING

8" x 8" BRACE

8" x 8" RANGER

3" x 8" PILING

12"

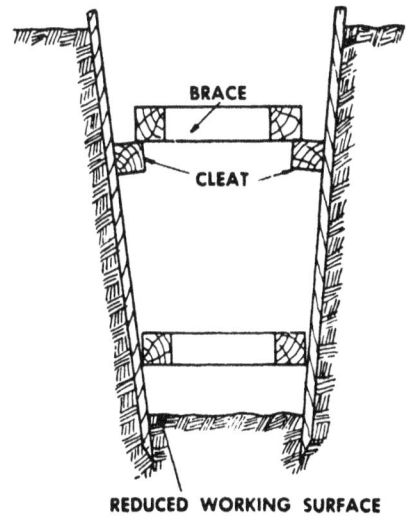

**EFFECT OF INCORRECTLY TURNED BEVEL IN DRYING SHEET PILING**

BRACE

CLEAT

REDUCED WORKING SURFACE

2   Trench-shoring methods

*Figure 4*—Continued.

# SECTION III
# FORMS AND REINFORCEMENT

## 9. General

Forms must be simple, economical, rigid, strong, practically watertight, and constructed so they can be readily removed without hammering or prying against the concrete. Form work should conform to specifications listed in appendix II, and the additional requirements in paragraphs below.

## 10. Design of Forms

Forms may be designed on the assumption that the hydrostatic pressures developed by green concrete against the form will be 150 pounds per square foot per foot of height up to a maximum of 1,000 pounds per square foot if the concrete is placed at a vertical rate of not more than 5 feet per hour.

## 11. Materials

Metal and wood are the materials most commonly used for forms in concrete work. Metal forms can be used repeatedly without additional cutting and fitting and they are often preferred. However, wood forms are satisfactory and are more generally used in ordinary building construction. The characteristics which should influence the selection of form materials, whether wood or metal, are: the strength needed to sustain the concrete load; weight of material to be hauled and handled; time required for preparing and erecting; possibility of reuse; and overall cost, including the cost of erecting and stripping.

a. *Wood.* Use any common grade lumber or exterior grade plywood for wood forms. If form lining is specified, build forms of square-edged boards; otherwise use either plywood or tongue-and-groove lumber of uniform width and thickness. Cut out open ring shakes, large checks, splits, loose knots, or other defects which might produce unsightly surfaces on the concrete; or discard the board in which such defects occur.

(1) *Green lumber.* Green lumber is suitable for form work because it has little tendency to swell and deform on contact with moisture.

(2) *Plywood.* Plywood form facing should comply with either Commercial Standards CS–45–55 (Douglas Fir) or CS 122–56 (Western Softwood), Grade A–C exterior, of thickness as required. Interior grade may be used for one-time use. Where a smooth finish is specified, plywood form facing is generally used in lieu of tongue-and-groove lumber, because of low labor costs in handling and high reuse value.

(3) *Sizes of lumber.* Form lumber must be selected to carry the anticipated load. However, the sizes of lumber commonly used in form work are 1-inch stock for floor, foundation, wall forms, columns, and beam sides; 2-inch stock for beam bottoms and heavy concrete construction; 2 x 4's for form studs, column yokes, and framing for panels; 2 x 6's or 2 x 8's for stringers & joists; 3 x 4's or 4 x 4's for posts, struts, shores, uprights, and sometimes stringers; 1- or 2-inch stock for cleats; 1 x 6's for cross ties and similar bracing.

b. *Composition Material.* Heavyweight paper or strawboard, impregnated with asphalt or oil and spirally wound into tubes of predetermined diameter, can be used as forms for piers of comparatively short lengths.

c. *Floor Forms for Combination Concrete and Hollow Tile or Metal Pan Construction.*

(1) Forms for these types of floors may be either "open" or "closed" deck construction (fig. 5). For closed deck construction, beam and girder forms and supporting members are built in the usual manner, with the entire floor area covered with 1-inch sheathing. With the open deck method for hollow tile, 2- by 8-inch planks are spaced 8 inches or more apart. This allows a tile bearing of about 2 inches and the usual 4-inch or more concrete joist. Hollow tile for floor construction should conform to Federal Specifications SS–T–321 (app. I).

(2) Because of the wide spacing of the deck planks, open deck construction is generally used with metal pans. Depending on the width of concrete joists, which may vary from 4 to 7 inches, metal pans (usually 20 to 30 inches wide) are set on 2- by 6-inch or 2- by 8-inch planks spaced 24 to 37 inches on centers.

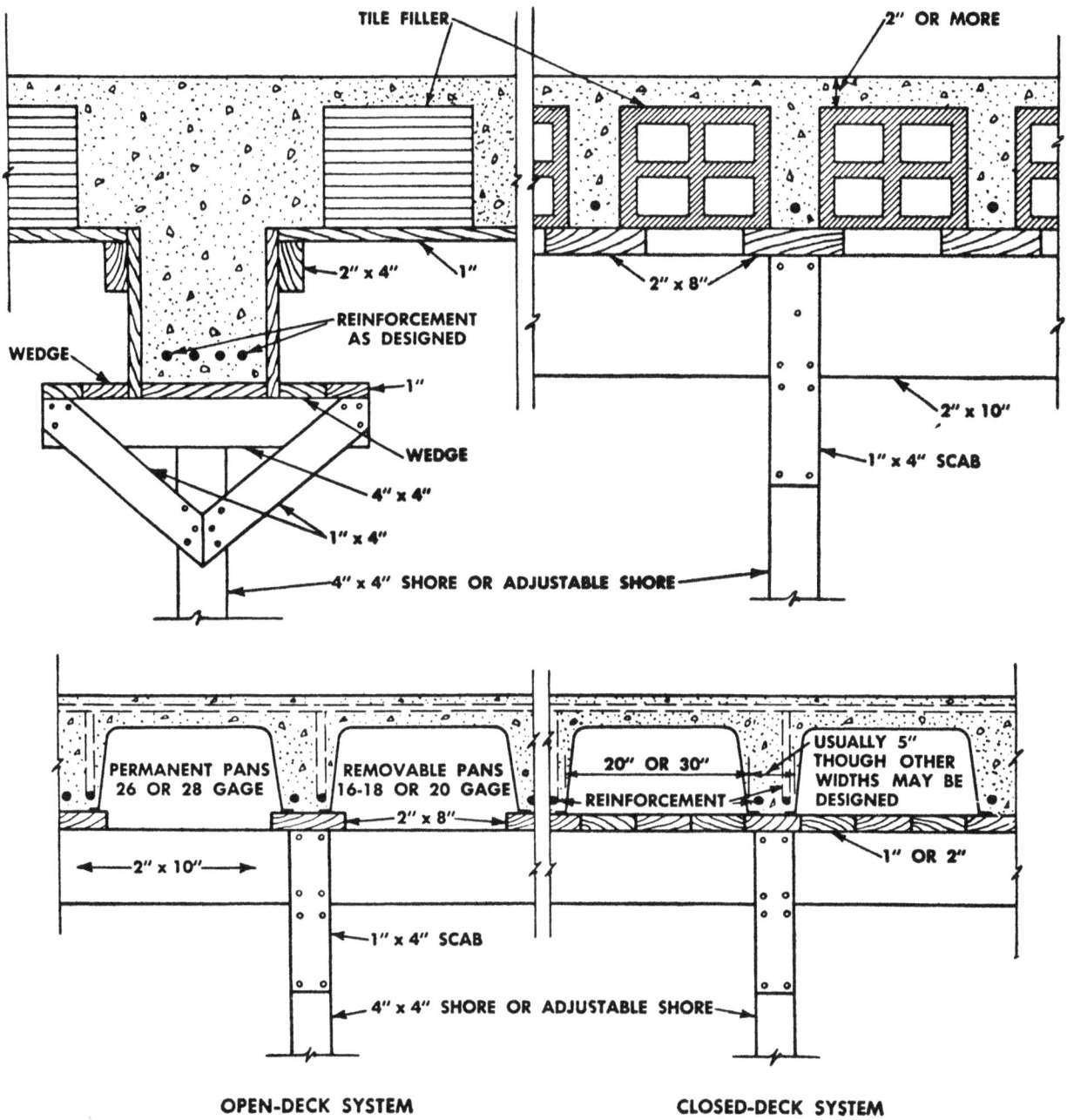

TILE FILLER

2" OR MORE

2" x 4"

1"

REINFORCEMENT
AS DESIGNED

WEDGE

1"

2" x 8"

WEDGE

4" x 4"

1" x 4"

2" x 10"

1" x 4" SCAB

4" x 4" SHORE OR ADJUSTABLE SHORE

PERMANENT PANS
26 OR 28 GAGE

REMOVABLE PANS
16-18 OR 20 GAGE

20" OR 30"

USUALLY 5"
THOUGH OTHER
WIDTHS MAY BE
DESIGNED

REINFORCEMENT

2" x 8"

1" OR 2"

2" x 10"

1" x 4" SCAB

4" x 4" SHORE OR ADJUSTABLE SHORE

OPEN-DECK SYSTEM

CLOSED-DECK SYSTEM

*Figure 5.  Forms for concrete floors.*

*d. Metal Pan Forms for Concrete-Joist Floors.*
  (1) Metal pan forms (fig. 6) are available in permanent and removable types. Sizes range from 20 to 30 inches wide, by 12, 24, 30, and 36 inches long, and 6, 8, 10, 12, and 14 inches deep. Filler pans for odd spaces are furnished 10, 12, 15, and 16 inches wide. End pans are made with one end closed and tapered.
  (2) Pans are made of both flat and corrugated metal sheets. Permanent pans (those left in place after the concrete has set) usually are made of 26- or 28-gage metal. Removable pans (those removed after the concrete has set) are made of 16-, 18-, or 20-gage metal, depending on the size of the pan and whether it is made of flat, corrugated, or ribbed sheets. It is not always economical to use pans of the lightest gage; they are easily damaged and frequent replacements may be necessary.
  (3) Tack pans to the wood forms securely so the pans remain rigid and in line while the reinforcing bars and concrete are placed.
  (4) Clean concrete and rust from pans removed from the slab and straighten dents and flanges. To prevent rusting, coat them with oil and store them under cover.
  (5) Use Metal-pan forms only for:
    (a) Areas where the slabs will be concealed by suspended ceilings and suitable floor finish.
    (b) Industrial-type facilities where appearance is not a major consideration.
  (6) For other areas, use flat-plate, two-way solid slab, beam and slab, or comparable type of structural floor or roof system.

## 12. Oil Coating

To insure removal of forms without spalling or roughening the concrete surface, oil those forms not having absorptive linings. Forms need not be oiled if the structure is to be faced with masonry or if backfill is to be used against it. Forms should not be oiled if concrete surface is to be painted or plastered. The kind of oil used depends on the type of form used and conditions under which concrete is poured.

*a. Wood Forms.* An oil suitable for wood forms penetrates the wood and leaves the surface oily enough to prevent the wood from sticking to the concrete or absorbing water and warping. Most light-colored and light-bodied straight petroleum oils sold as form oils are acceptable. (See app. II) Blends of fuel oil with kerosene or paraffin and crankcase oil may stain the concrete. Dark oils and oils containing excessive free carbon or sludge are not satisfactory for use on wood forms.

*b. Steel Forms.* Oils satisfactory for wood forms are often unsatisfactory for steel. Compound oils are more effective than straight petroleum oils in resisting abrasion and in preventing concrete from sticking to steel. Compound oils are composed principally of petroleum oil, but also contain other oils of animal or vegetable origin, gums, and resins; they are heavier in body and darker in color than straight petroleum oils.

## 13. Wetting Wood Forms

All wood forms not otherwise treated should be wetted before concrete is poured. In freezing weather, wetting can be omitted if steam jets are used to remove frost and ice coatings.

## 14. Forming

*a. Building and Anchoring.* Be sure that forms are built true to the lines and grades shown on the drawings. Set boards close together to make joints mortartight. Anchor forms securely with wedges, bolts, braces, or other means, to prevent distortion or buckling before the concrete sets.

  (1) *Forms for steps supported on ground.* Construct as illustrated and described in 1, figure 7.
  (2) *Forms for self-supporting reinforced steps.* Construct as illustrated in 2, figure 7. (Nail side planks to panel and strengthen with outside braces.)

*b. Erection.* Erect the controlling face panel, usually the outside one, and bring it true to line and plumb; then build the other form face and hold it parallel to the first by spreaders as long as the desired wall thickness. Twisted, annealed, black-wire ties and 1 x 2 wood spreaders (fig. 8) are the most commonly used and generally the least expensive. Do not use wire ties if rust stains will be objectionable in the finished concrete surface. Several patented metal bolt spreader-and-tie combinations (fig. 8) may be used on large projects where form faces must be accurate and rust stains are objectionable.

*c. Centering Columns and Piers.* Center the forms for columns and piers carefully. Install chamfer

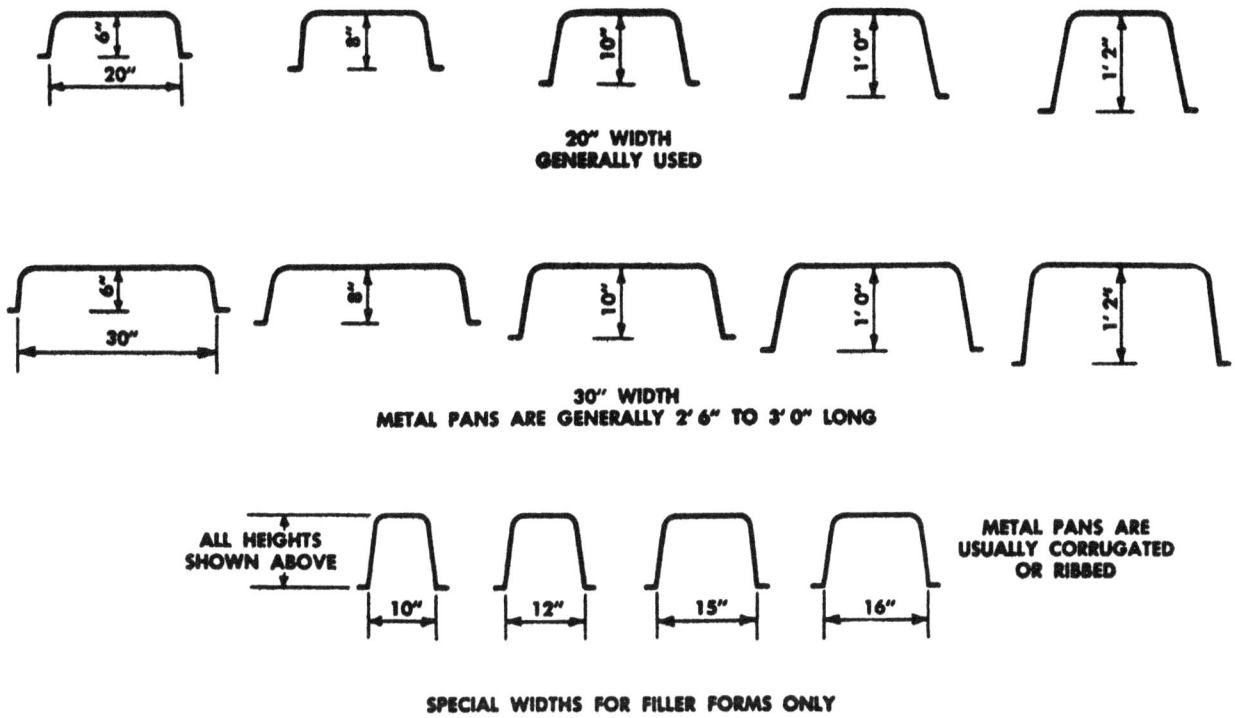

6"

20"

8"

10"

1' 0"

1' 2"

**20" WIDTH
GENERALLY USED**

6"

30"

8"

10"

1' 0"

1' 2"

**30" WIDTH
METAL PANS ARE GENERALLY 2' 6" TO 3' 0" LONG**

ALL HEIGHTS
SHOWN ABOVE

10"

12"

15"

16"

**METAL PANS ARE
USUALLY CORRUGATED
OR RIBBED**

**SPECIAL WIDTHS FOR FILLER FORMS ONLY**

*Figure 6.    Metal pan forms.*

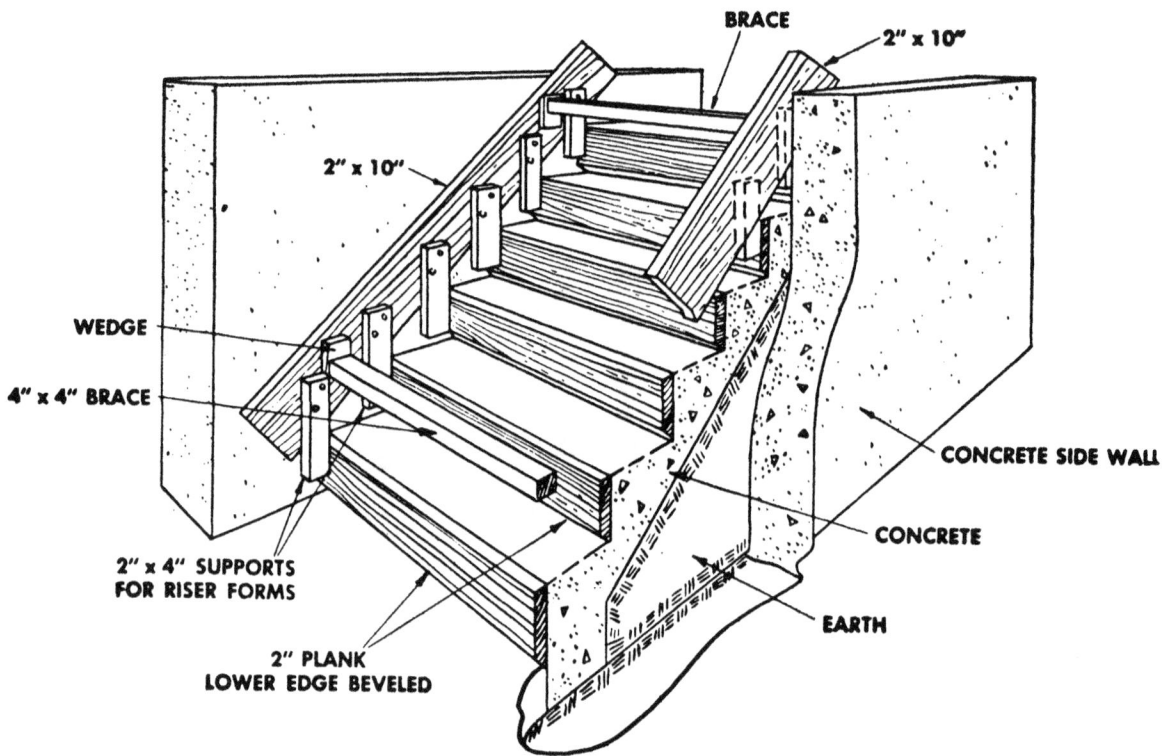

BRACE

2" x 10"

2" x 10"

WEDGE

4" x 4" BRACE

CONCRETE SIDE WALL

CONCRETE

EARTH

2" x 4" SUPPORTS
FOR RISER FORMS

2" PLANK
LOWER EDGE BEVELED

①

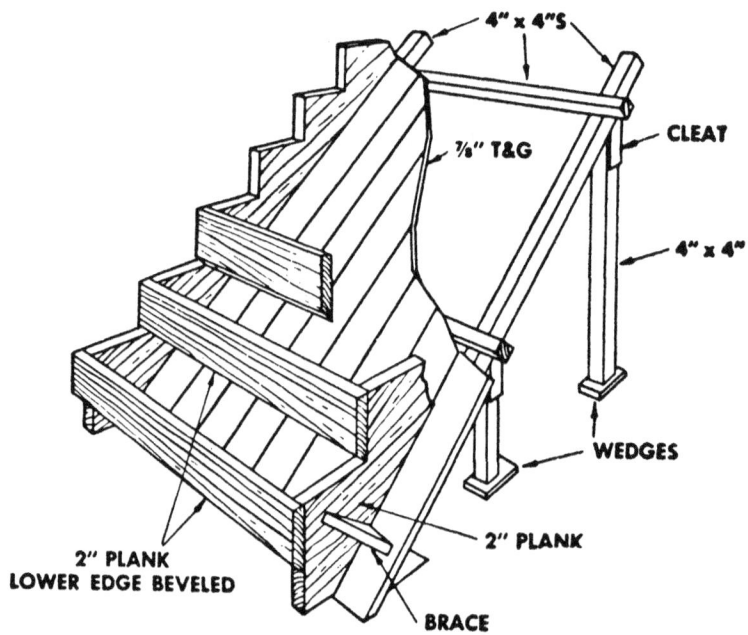

4" x 4"S

⅞" T&G

CLEAT

4" x 4"

WEDGES

2" PLANK

2" PLANK
LOWER EDGE BEVELED

BRACE

②

1  Forms for steps supported on ground        2  Form for self-supportign reinforced steps

*Figure 7.  Forms for concrete steps.*

AGO 10179A

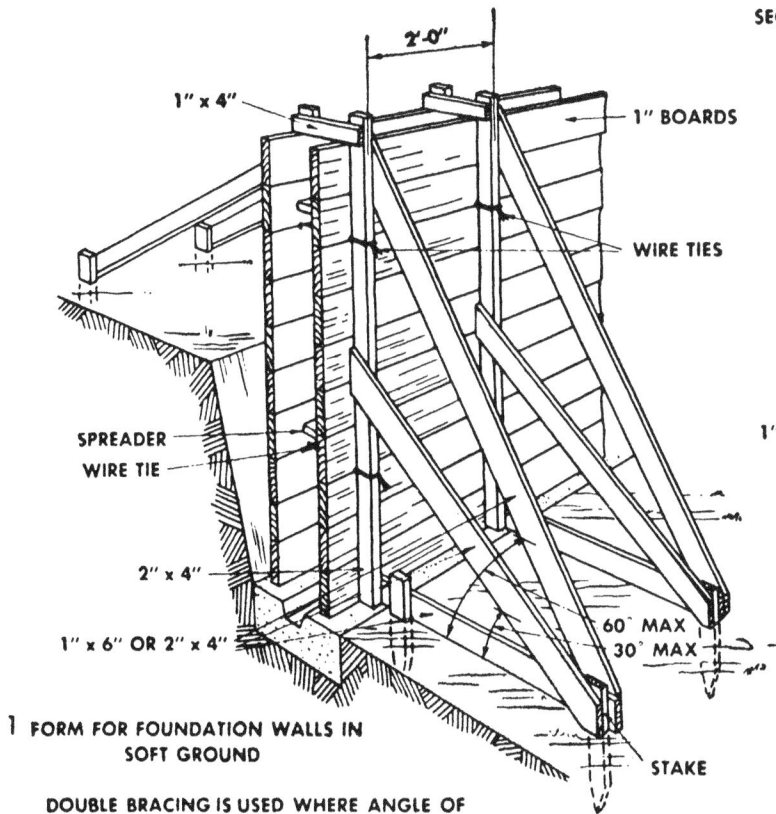

**1** FORM FOR FOUNDATION WALLS IN SOFT GROUND

DOUBLE BRACING IS USED WHERE ANGLE OF SINGLE BRACE IS MORE THAN 45.

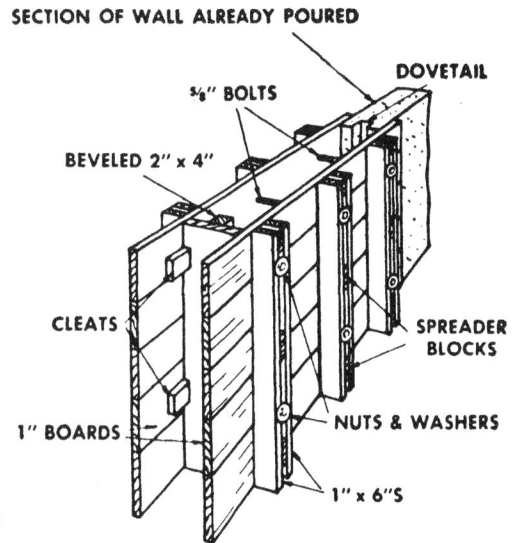

**3** FORM FOR WALL TO BE CAST IN SECTIONS

NOTE DOVETAIL FORMED IN WALL ALREADY POURED FOR PURPOSE OF HOLDING SECTIONS OF WALL IN LINE. BOLTS USED INSTEAD OF WIRES.

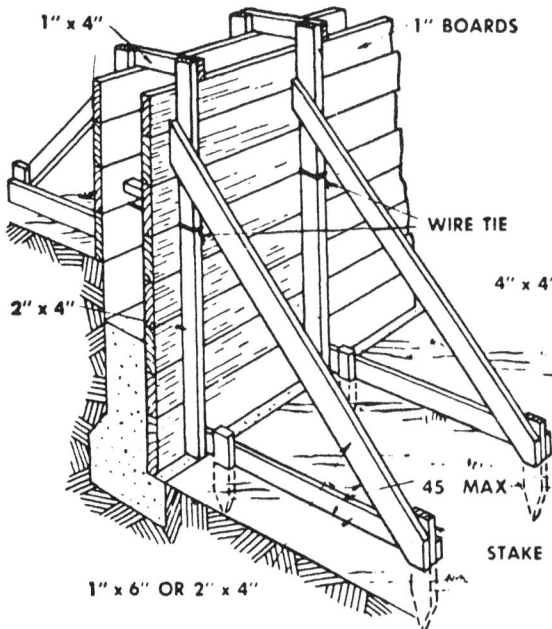

**2** FORM FOR WALL IN FIRM GROUND

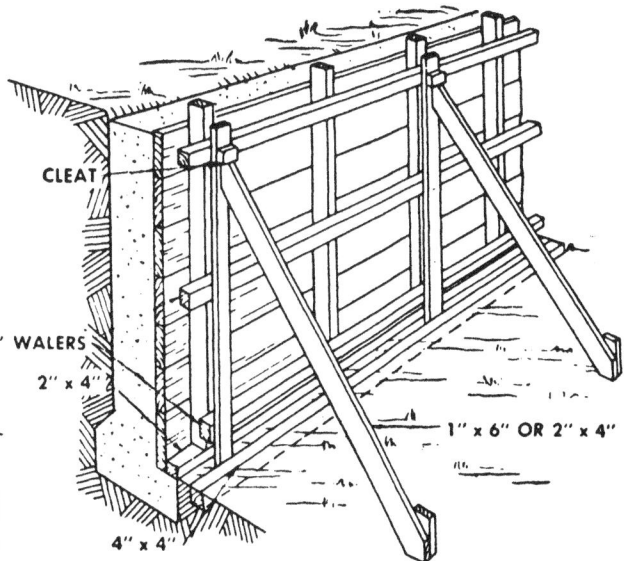

**4** FORM USING WALERS

1  Form for foundation wall in soft ground  2  Form for wall in firm ground  3  Form for wall to be cast in sections  4  Form using walers

Figure 8.  Forms for concrete walls.

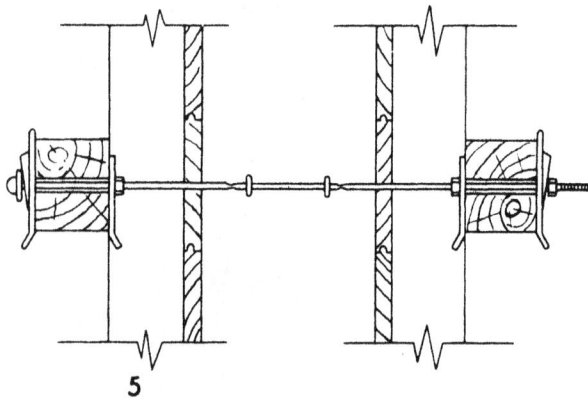

5

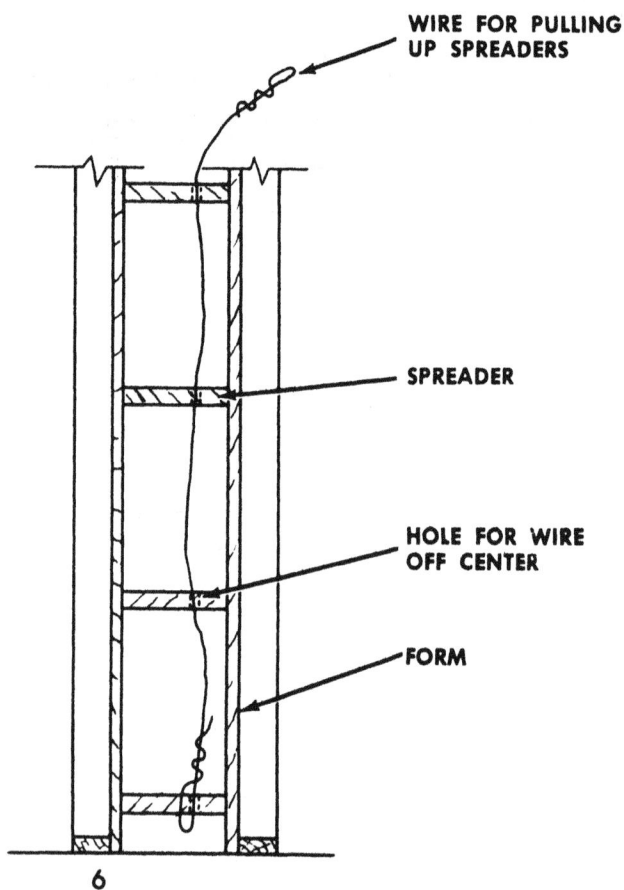

WIRE FOR PULLING
UP SPREADERS

SPREADER

HOLE FOR WIRE
OFF CENTER

FORM

6

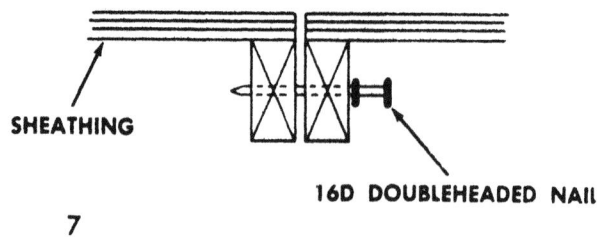

SHEATHING

16D DOUBLEHEADED NAIL

7

| 5 Tie rod and spreader for wall form | 6 Method for the removal of wood spreaders | 7 Method of connecting wall form panels together |

*Figure 8*—Continued.

strips in corners of column forms to facilitate stripping forms and to produce columns without sharp corners. Wood yokes or clamps (fig. 9) speed up construction and tie forms together tightly. Several types of adjustable metal column clamps are available for repeated use.

d. *Supporting Shores.*

    (1) Support forms for slabs, beams, and girders (which are to be poured in place) by adjustable jack shores or built-in timber shores. Compute total load of wet concrete; use the number and size of shores

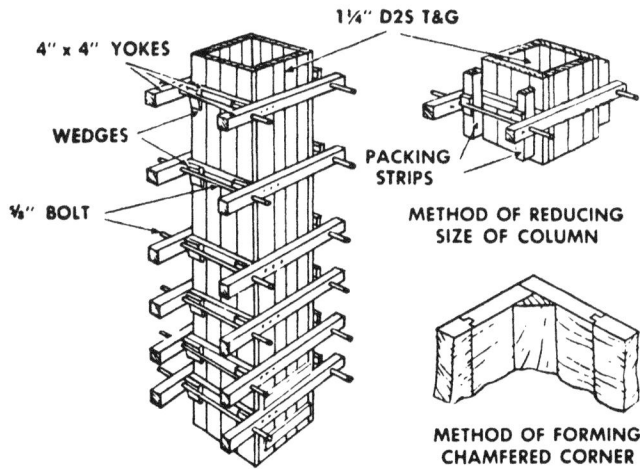

WOOD FORMS FOR CONCRETE COLUMNS

METHOD OF REDUCING SIZE OF COLUMN

METHOD OF FORMING CHAMFERED CORNER

| SPACING OF YOKES FOR COLUMNS | | | | | |
|---|---|---|---|---|---|
| | LARGEST DIMENSION | | | | |
| HEIGHT | 16" | 18" | 20" | 24" | 28" |
| 1' | | | | | |
| 2' | 31" | 29" | 27" | 23" | 21" |
| 3' | | | | 23" | 21" |
| 4' | 31" | 28" | 26" | 23" | 21" |
| 5' | | | | 23" | 20" |
| 6' | | 28" | 26" | 23" | |
| 7' | 30" | | | 22" | 18" |
| 8' | | 26" | 24" | 16" | 15" |
| 9' | | | | | 13" |
| 10' | 29" | | 19" | 14" | 12" |
| 11' | | 20" | 16" | 13" | 9'9" |
| 12' | | 18" | 15" | 12" | 9'9" |
| 13' | 21" | | 14" | 11" | 9'9" |
| 14' | 20" | 16" | | 1'0" | 8'8" |

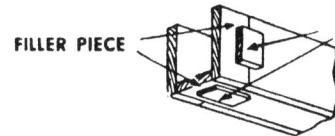

FILLER PIECE

METHOD OF LENGTHENING GIRDER OR BEAM FOR SMALLER COLUMN

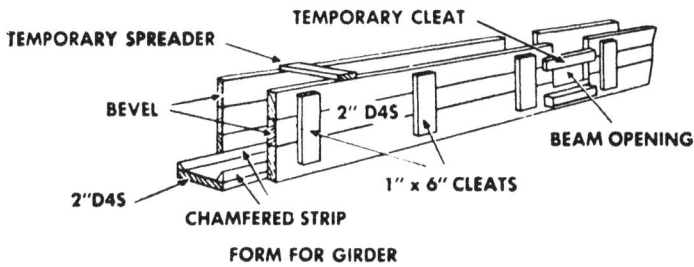

TEMPORARY SPREADER

TEMPORARY CLEAT

BEVEL

2" D4S

BEAM OPENING

2" D4S

1" x 6" CLEATS

CHAMFERED STRIP

FORM FOR GIRDER

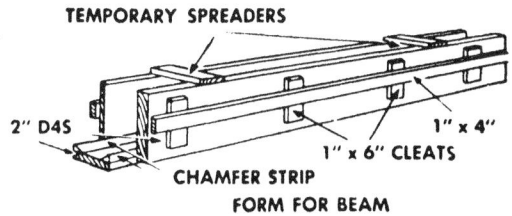

TEMPORARY SPREADERS

2" D4S

1" x 4"

1" x 6" CLEATS

CHAMFER STRIP

FORM FOR BEAM

GIRDER SIDE

BEAM SIDE

DETAIL AT A

3/8"

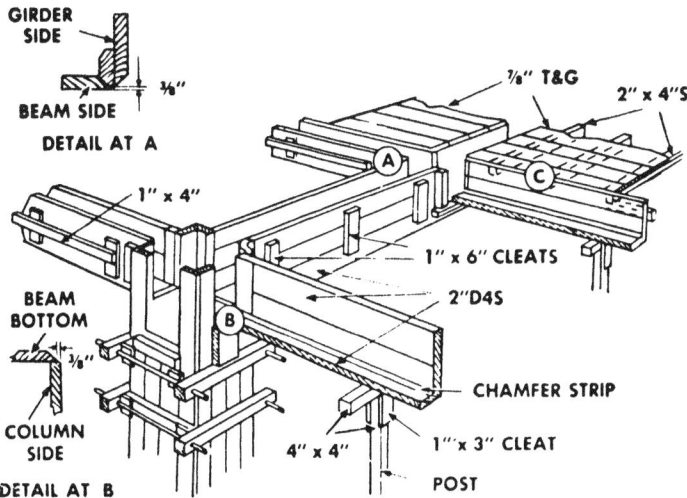

7/8" T&G

2" x 4"S

1" x 4"

1" x 6" CLEATS

2" D4S

CHAMFER STRIP

1" x 3" CLEAT

4" x 4"

POST

BEAM BOTTOM

3/8"

COLUMN SIDE

DETAIL AT B

COLUMN GIRDER BEAM & SLAB FORMS ASSEMBLED

BEVELED EDGE

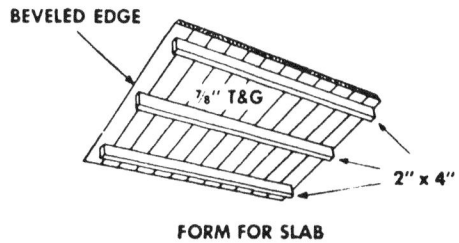

7/8" T&G

2" x 4"

FORM FOR SLAB

3/8"

2" x 4"

CLEAT

1" x 4"

DETAIL AT C

Figure 9. Forms for concrete members.

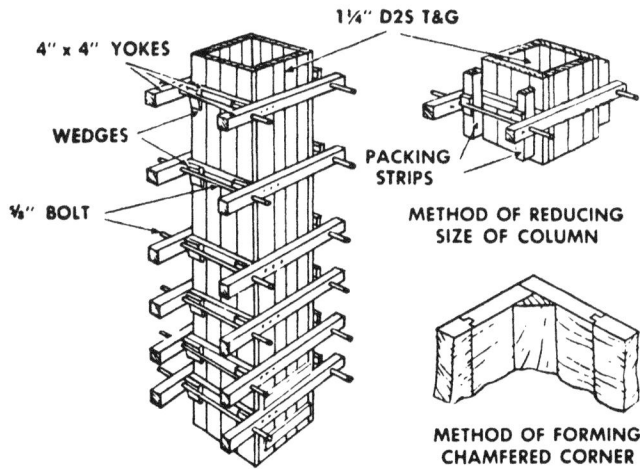
4" x 4" YOKES
WEDGES
1/2" BOLT
1 1/4" D2S T&G
PACKING STRIPS

2" x 10" PLANK FLOOR

4" x 4"

S 1" x 6" LEDGER

2" x 4"

2" x 4"

4" x 4"

2" x 4"

2" x 4"

2" x 4"

WHEEL BARROW RUNWAY

5—2" x 10" RUNWAY PLANKS

2" x 4"

TO MIXER →

1" x 6"

2" x 6" RUNWAY SUPPORT

4" x 4"

1" x 6" CROSS BRACE

2" x 4"

2" x 6" FOOTING

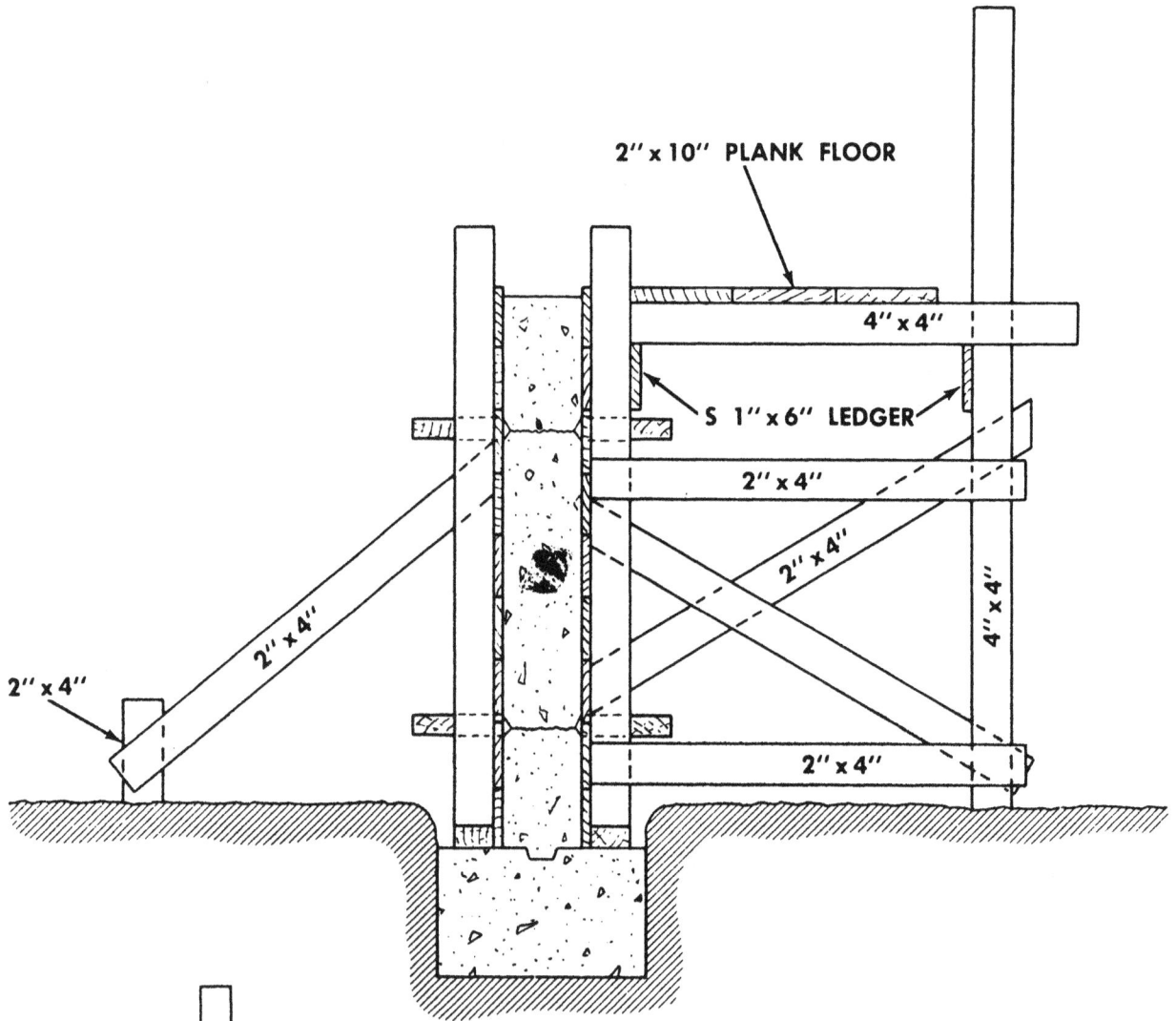

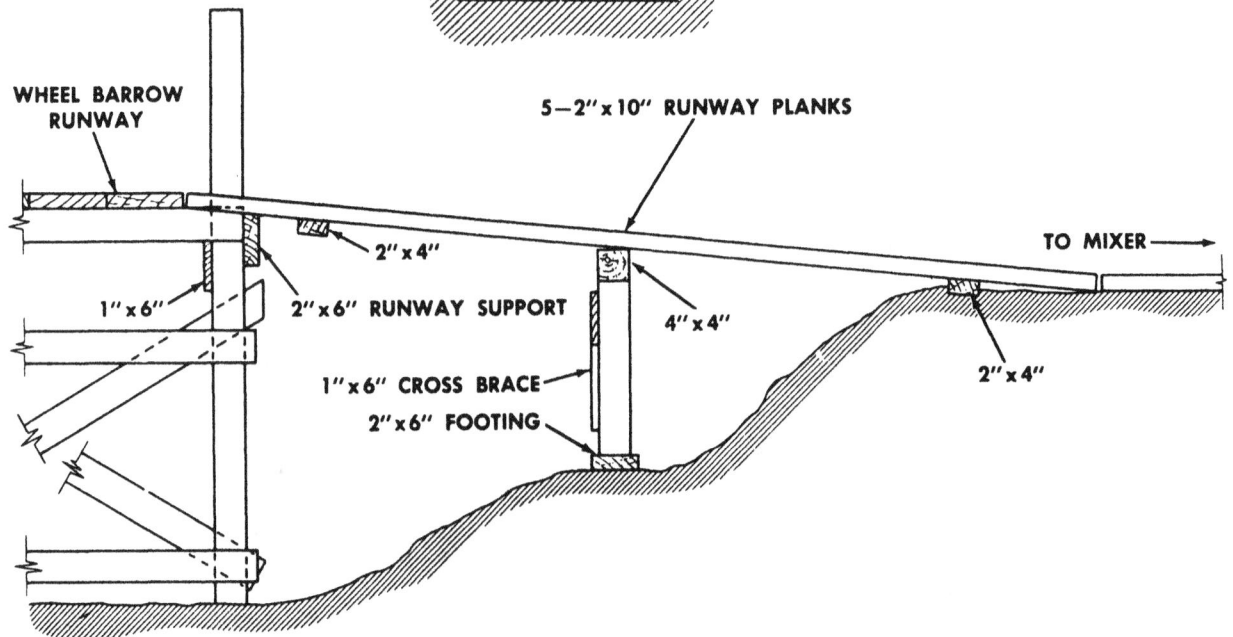

1   Runway for placing concrete by buggy or wheelbarrow                    2   Inclined runway

*Figure 10.   Runways.*

2-2 x 12

1 x 5

2 x 8

THIS DIAGONAL BRACE OR
THE BRACE ON THE OPPOSITE
SIDE MAY BE REMOVED TO
PERMIT ACCESS TO THE TOWER

4 x 4 POST

ELEVATOR

GUIDES

PULLEY ARRANGEMENT

2 x 8

GUIDES

ELEVATOR FOR MATERIAL TOWER

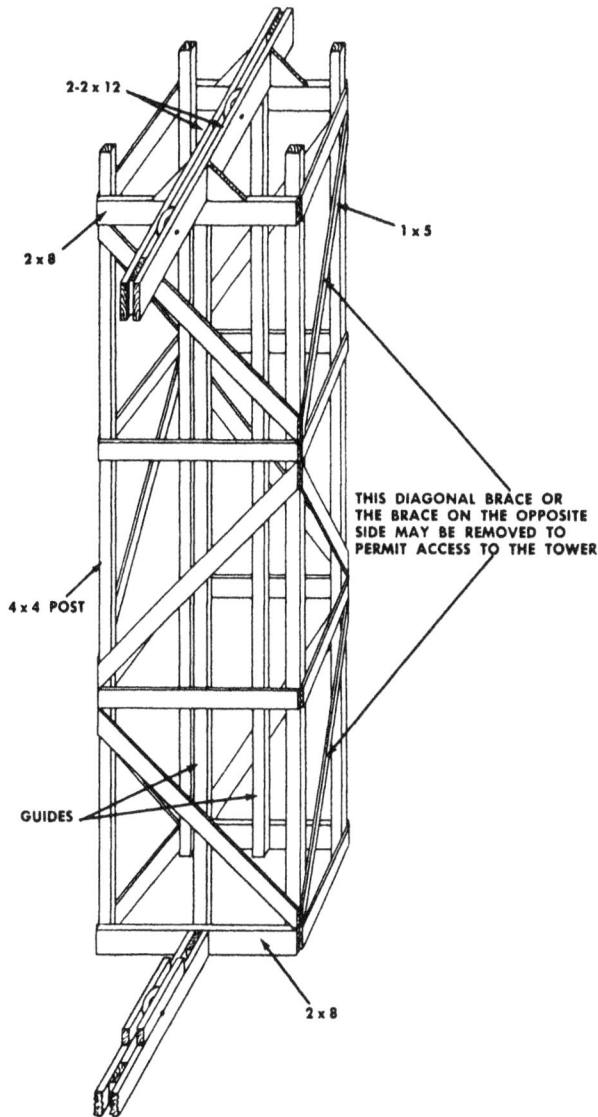

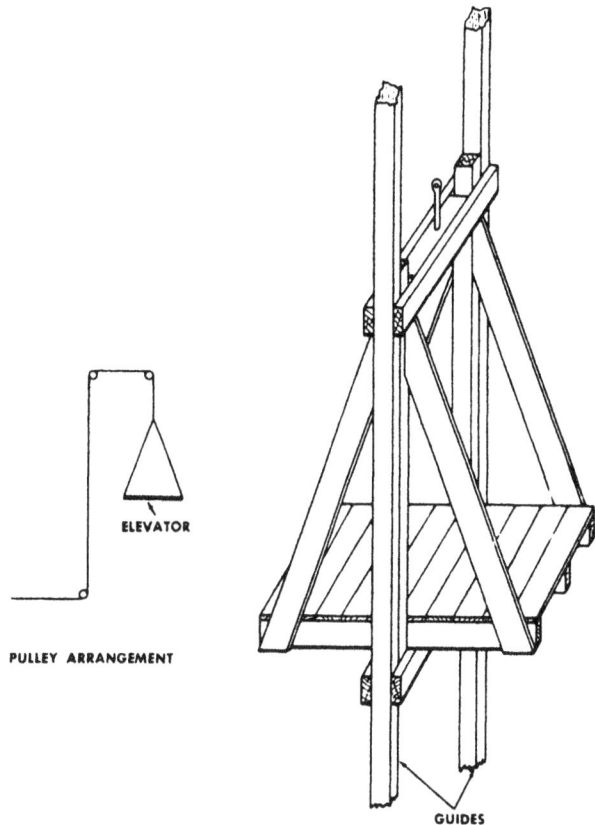

1   Material tower                                    2   Elevator   for   material   tower

*Figure 11.   Material tower and elevator.*

needed to carry the load safely. Give beam and girder forms a camber sufficient to prevent excess sag of finished members when forms are removed and loads applied.

(2) Adjustable shores are usually combinations of wood and metal made in several sizes, which can be raised or lowered within certain limits. Adjustable shores are easier to put in place than all-timber shores which require wedging, cutting, or adding to for each change in story height. The original cost of adjustable shores is greater than that of timber used for the same purpose;

however, their use results in saving because of reduced labor costs.

*e. Runways and Hoists.* Make arrangements for moving materials for every concrete pour.

(1) For most small and medium sized projects, construct simple ramps and runways as shown in figure 10. The concrete mixing operations and ramp to the pour should be located on terrain that will permit the shortest and least slope on ramps.

(2) Provide elevators or skips for large projects (fig. 11).

*f. Form Linings.* For finer appearance of walls and ceilings exposed to view, line forms with large sheets of moisture-resistant plywood or an equivalent pressed wood lining material. Use full-length pieces with close joints to line window and door frames, and their flat or arched soffits. Do not use patched pieces, or linings with bruises, hammer marks, or other defects which will show in the finished concrete surface.

*g. Inspection.* Before placing concrete, check forms to insure:

(1) Clean-out openings for removal of debris are provided at foot of forms.
(2) Debris, water, ice, and snow have been removed from forms.
(3) Brick cut-offs, pipe chases and sleeves, liners, inserts for hangers, window and door frames, and the like are properly installed.
(4) Bulkheads are installed at open ends of walls and places where wall elevation drops.
(5) Forms have enough bracing to prevent bulging, distortion, or collapse.
(6) Steel dowels and anchor bolts are held securely in place by suitable templates.
(7) Size and spacing of reinforcing bars are as shown on drawings.
(8) Tie wires for reinforcing bars are strong enough to hold them in place while concrete is being poured, (fig. 12).

## 15. Removal of Forms

*a. General.* Time required for concrete members to reach safe strengths varies with type of cement used, concrete mixture, atmospheric temperature during pouring and curing, and type of member placed. Usually, concrete forms should remain in place longer for reinforced concrete than for plain mass concrete. Horizontal members (beams and slabs) need more time than vertical members (columns and walls) to cure and attain strength to support their own weights plus imposed loads.

(1) Do not remove forms until concrete has hardened adequately. Do not remove shoring under floor slabs and roof decks until the supported concrete member can safely support its own weight plus any imposed load.
(2) Twenty-four hours after concrete is placed, loosen tie rods which are to be entirely removed. At the same time, remove ties not needed to hold forms in place, pulling them toward the inside face of the wall.

(3) When removing forms, take care to avoid spalling the concrete surface.
(4) To prevent termite infestation, remove all wood, impregnated strawboard, and similar materials from piers, walls, and under surfaces of porch floors, steps and similar structural members.
(5) Withdraw all nails as forms are stripped from concrete.

*b. Time Schedule.* When type I portland cement meeting Federal Specification SS-C-192 is used in concrete mixtures under normal weather conditions, follow the schedule below for safe form removal.

| Concrete items | Above 60° F. | 50° F. to 60° F. | 40° F. to 50° F. |
|---|---|---|---|
| Footing forms | 2 days | 5 days | 7 days |
| Column and wall forms and side forms of beams and girders. | 5 days | 7 days | 10 days |
| Bottom forms of slabs, 6-foot span or less. Add 12 hours for each additional foot of span. | 5 days | 9 days | 14 days |
| Bottom forms of beams and girders, 14-foot span or less. Add 24 hours for each additional foot of span. | 14 days | 18 days | 31 days |

When temperature is below 40°F., keep forms and shores in place an additional time equal to the time structure has been exposed to lower temperature.

## 16. Waterstops

To prevent the passage of water provide waterstops in joints where necessary. Waterstops may be on non-ferrous metal, rubber, or plastic; install so as to form a watertight joint. Metallic waterstops should normally be of 20-ounce copper; plastic waterstops should normally be of polyvinyl chloride, or other such material of proven performance.

## 17. Reinforcing Steel

*a. General.* All steel requirements for reinforced concrete, unless otherwise specified, should conform to the American Concrete Institute Manual of Standard Practice for Detailing Reinforced Concrete Structures (ACI-315), and their Building Code Requirements for Reinforced Concrete (ACI-318). Show sufficient information on the drawings to permit rapid preparation of necessary bar lists, bending diagrams, and shop drawings.

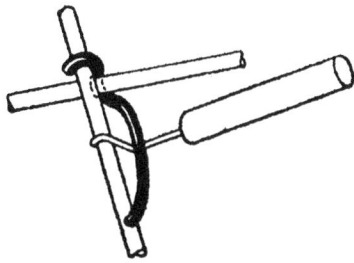

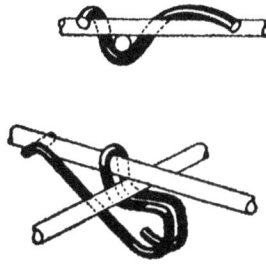

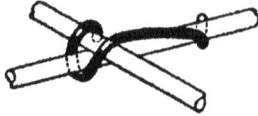

FASTENERS AND TIES

1" MIN, WALLS, FIREPROOFING
2" MIN, WALLS TO WEATHER

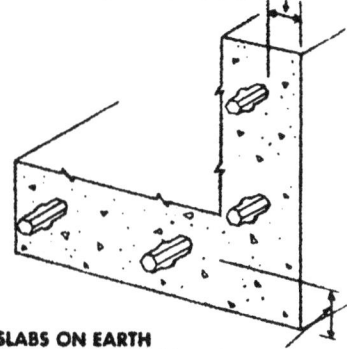

3" MIN, SLABS ON EARTH
1" MIN, SLABS OR FIREPROOFING
1½" MIN, BEAMS, GIRDERS, COLUMNS

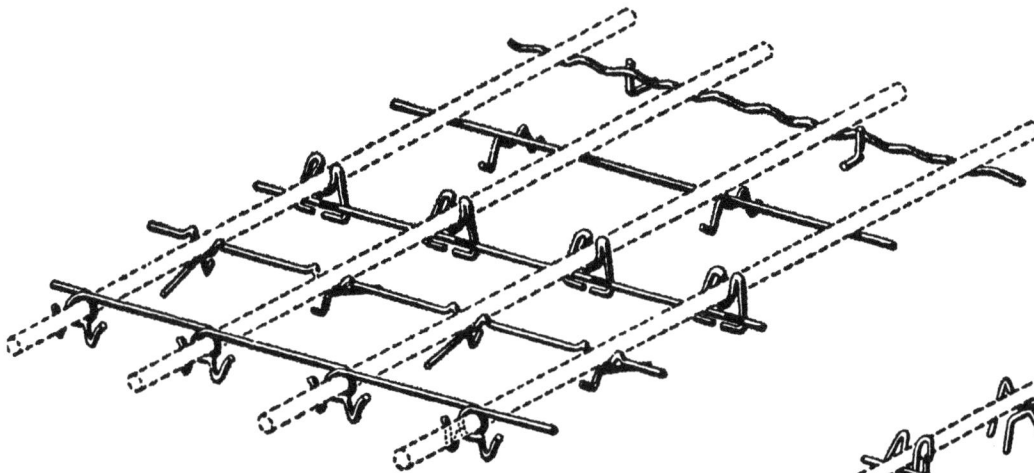

CONTINUOUS BAR SUPPORTS

NOT TO SCALE

LAP AT LEAST 40 x BAR DIAMETER

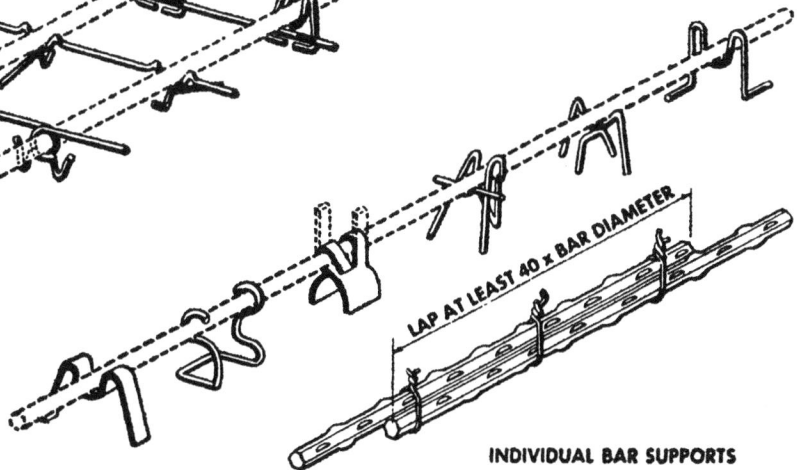

INDIVIDUAL BAR SUPPORTS

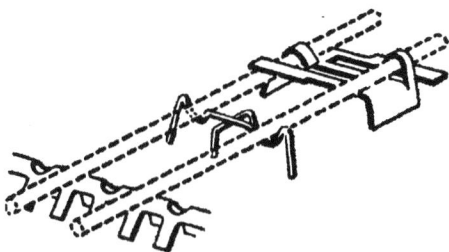

BEAM AND JOIST BAR SUPPORTS

| ANNEALED-IRON WIRE A S &W GAGE | | | | | | |
|---|---|---|---|---|---|---|
| GAGE NO | DIAMETER (IN) | | FT PER LB | AREA (SQ IN) | BREAKING STRENGTH (LBS) | SAFE LOAD (LBS) |
| | DECIMAL | FRACTION | | | | |
| 4 | 0.225 | 15/64 | 7 | 0.040 | 2500 | 625 |
| 6 | 0.192 | 13/64 | 10 | 0.028 | 1800 | 450 |
| 8 | 0.162 | 11/64 | 14 | 0.020 | 1300 | 325 |
| 10 | 0.135 | 9/64 | 20 | 0.013 | 900 | 225 |
| 12 | 0.105 | 7/64 | 33 | 0.008 | 600 | 150 |
| 14 | 0.080 | 5/64 | 58 | 0.005 | 350 | 88 |
| 16 | 0.062 | 1/16 | 96 | 0.003 | 220 | 55 |
| 18 | 0.047 | 3/64 | 166 | 0.002 | 130 | 32 |

*Figure 12. Reinforcement fasteners and supports.*

*b. Quality of Materials.* Reinforcement bars should conform to Federal Specifications QQ–S–632, Type II, Grade C, D, E, or G. Wire fabric reinforcement should conform to the requirements of ASTM Designation A 185, Welded Steel Fabric for Concrete Reinforcement, and be of the wire and mesh sizes specified or shown on the drawings.

*c. Bending.* Reinforcing bars, stirrups, spacer rods, and similar secondary reinforcement must meet the bend-test requirements of Federal Specification QQ–S–632 and be accurately mill-bent or field-bent to forms and dimensions indicated on the general drawings, bending diagrams, or as approved by the designing engineer.

(1) Bars for test must be bent cold around a pin without cracking, with the degree of the bend and size of pin as specified in bend-test requirements of QQ–S–632.

(2) Check each shipment, especially mill-bent bars, for damage in transit.

(3) Do not straighten bars or readjust bends in any manner that may cause fracture.

(4) Reject bars which rupture during bending operations.

(5) Bend bars cold; avoid heating to facilitate bending.

*d. Minimum Spacing of Bars.* Unless specifically authorized otherwise, the clear space between parallel bars is never less than 1 inch. Normally, the clear space should be not less than the diameter of round bars, $1\frac{1}{2}$ times the side dimension of square bars, nor $1\frac{1}{3}$ times maximum size of coarse aggregate.

*e. Minimum Concrete Covering.* The design should provide enough concrete covering to protect the reinforcement from possible damage from frost, fire, moisture, gases, or other hazards. Minimum concrete covering for reinforcement is given below. Dimensions are measured from face of reinforcement to nearest face of forms.

(1) Main reinforcement in sewer conduits, culverts, walls and footings, and similar items of concrete embedded in earth, submerged in or exposed to the action of water alkali or gases, or subject to severe impacts or abrasions, must have a covering of at least 3 inches.

(2) Main reinforcement in beams, slabs, and walls of culverts, bridges, and similar items or comparatively thin sections exposed to weather, must have a covering of at least $1\frac{1}{2}$ to 2 inches.

(3) Main reinforcement in walls and slabs exposed to the weather, and in fire-resistant construction, must have a covering of—

(a) At least 1 inch for walls and slabs.

(b) At least $1\frac{1}{2}$ inches for floor beams.

(c) At least 2 inches in girders and columns.

(4) In interior flat-slab construction, the minimum slab covering may be reduced to $\frac{3}{4}$ inch.

(5) For interior construction where fire hazards do not exist, the covering over main reinforcement can be reduced to:

(a) Three-fourths inch in walls and slabs.

(b) One inch in floor beams.

(c) One and one-half inches in girders and columns.

(6) The covering over rods for stirrups, spacers, and similar secondary reinforcement can be reduced by the diameter of such rods.

*f. Splicing.* In general, indicate position and length of all reinforcement splices on the drawings. In designing, make sure that unsupported lengths of vertical reinforcement are not excessive and that splices in main reinforcement do not occur at points of maximum moment. Lap bars at splices at least 40 diameters or in accordance with table in ACI 315 and secure splicing so as to transfer the stress by bond. Stagger splices in parallel bars. Adjacent sheets of reinforcing mesh are spliced by lapping not less than 6 inches or one mesh. Lap ends of mesh sheets and wire them securely.

*g. Supports.* Assure that metal or concrete supports, spacers and/or ties, are sufficiently strong to hold reinforcement in place throughout the concreting operation. Supports should not be exposed in the finished concrete surface.

*h. Protection.* Assure that when concrete is placed, bar and mesh reinforcement and accessories are entirely free from dirt, rust, scale, grease, or other coatings which will destroy or reduce bond. Protect exposed reinforcement intended for bonding with future work by heavy wrappings of burlap saturated with bituminous material. Clean reinforcement so protected before concrete is placed.

# SECTION IV
# PREPARING AND PLACING CONCRETE

## 18. General

Concrete maintenance and repair are largely problems of small-scale construction, cutting, patching, and waterproofing. Concrete work should be done in accordance with appendix II and any additional requirements in this section. Concrete should be composed of portland cement, water, fine and coarse aggregate, and where needed entrained air. Select and control the proportions of the concrete mixtures to produce the quality of concrete appropriate for the intended service.

## 19. Materials

Production of satisfactory concrete requires good materials correctly graded, mixed, placed, and finished.

*a. Cement.* Portland cement should conform to the requirements of Federal Specification SS–C–192 and be of the following types for conditions prevalent:

Type I—For use in general concrete construction when the special properties of other types are not required.

Type II—For use in general concrete construction and in construction exposed to moderate sulfate action, or where moderate heat of hydration is required.

Type III—For use where high early strength is required.

Type IV—For use where low heat of hydration is required.

Type V—For use where high sulfate resistance is required.

Types 1–A, II–A, and III–A are the same as corresponding types above except that these contain an air entraining addition. They should be used for concrete construction and repair work exposed to the weather, unless an air-entraining admixture is to be added to the concrete when the materials are batched for mixing. Use high-early-strength cement when circumstances do not permit curing for the normal length of time, or work is to be placed in service at the earliest possible date. If the possibility of sulfate attack exists, as from alkaline surface, ground water, or sea water, use Types II or V.

*b. Aggregates.* For allowable concentrations of foreign matter, (shale, clay lumps, silt, or dust, and of soft, friable, thin, flaky, elongated, or laminated particles) see Federal Specifications SS–A–281.

(1) *Silt content.* Determine the silt content of sand by the quart-jar method. Put 2 inches of sand in a quart jar and add water until the jar is about three-quarters full. Shake vigorously for 1 minute and allow to stand for 1 hour. Silt content is too great if more than $\frac{1}{8}$-inch layer of silt forms on surface of sand (fig. 13). Discard sand containing excessive silt or remove silt by washing. Use specially equipped plants to wash silt out of large quantities of sand and gravel. Wash small batches on the job, using a device like that shown in figure 14. Figure 15 shows ideal sand gradation.

(2) *Organic materials.* Do not use aggregates containing harmful organic materials. Determine presence of these materials by using the American Society for Testing Materials designation C 40, "Test for Organic Impurities in Sand."

*c. Water.* Use fresh, clean water, suitable for drinking (potable) where practicable.

*d. Admixtures.*

(1) *General.*

(a) Concrete can be greatly improved by adding chemical compounds and other substances during mixing operations. Improvements most generally sought are:

1. Improved workability at reduced water-cement ratio.
2. Reduced finishing effort.
3. Reduced segregation of materials.
4. Hastened or retarded setting time.
5. Increased strength.
6. Denser and more waterproof concrete.
7. Reduced "bleeding" (rising of water to surface).
8. Increased durability.
9. Reduced cost.

(b) There is a strong relationship between most of the above characteristics. For instance, if workability is increased and water cement ratio decreased also, it is axiomatic that segregation is reduced, durability is increased, concrete is made

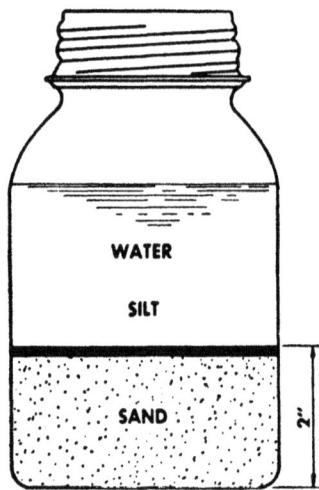

*Figure 13. Quart-jar method of determining silt content of sand.*

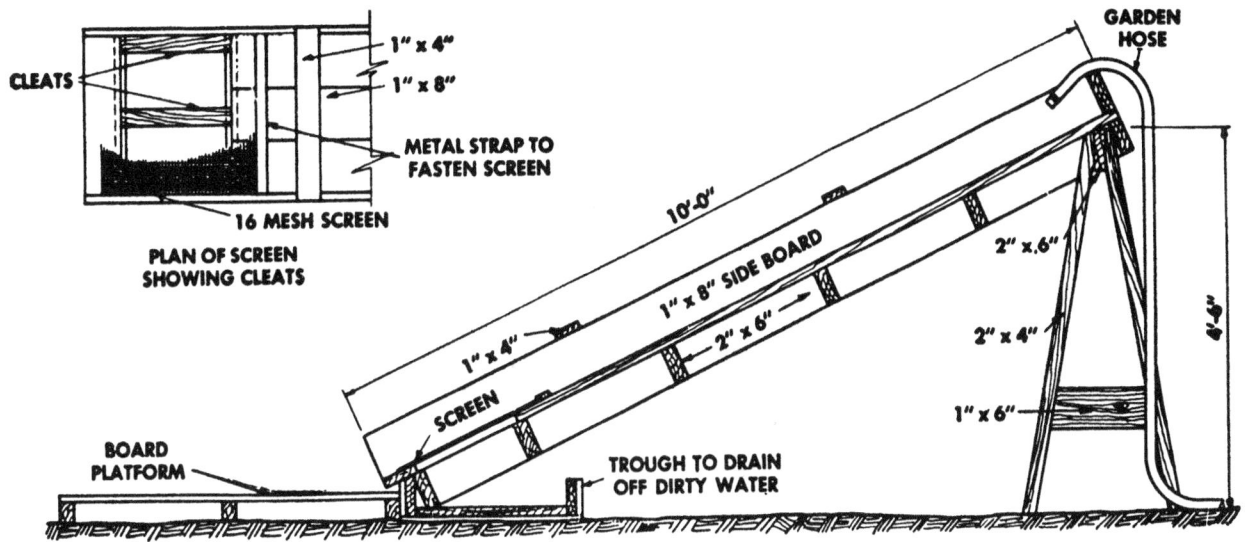

CLEATS

1" x 4"

1" x 8"

METAL STRAP TO
FASTEN SCREEN

16 MESH SCREEN

PLAN OF SCREEN
SHOWING CLEATS

GARDEN
HOSE

10'-0"

1" x 8" SIDE BOARD

1" x 4"

2" x 6"

SCREEN

2" x,6"

2" x 4"

1" x 6"

4'-4"

BOARD
PLATFORM

TROUGH TO DRAIN
OFF DIRTY WATER

*Figure 14. Device for washing sand and gravel with high silt content.*

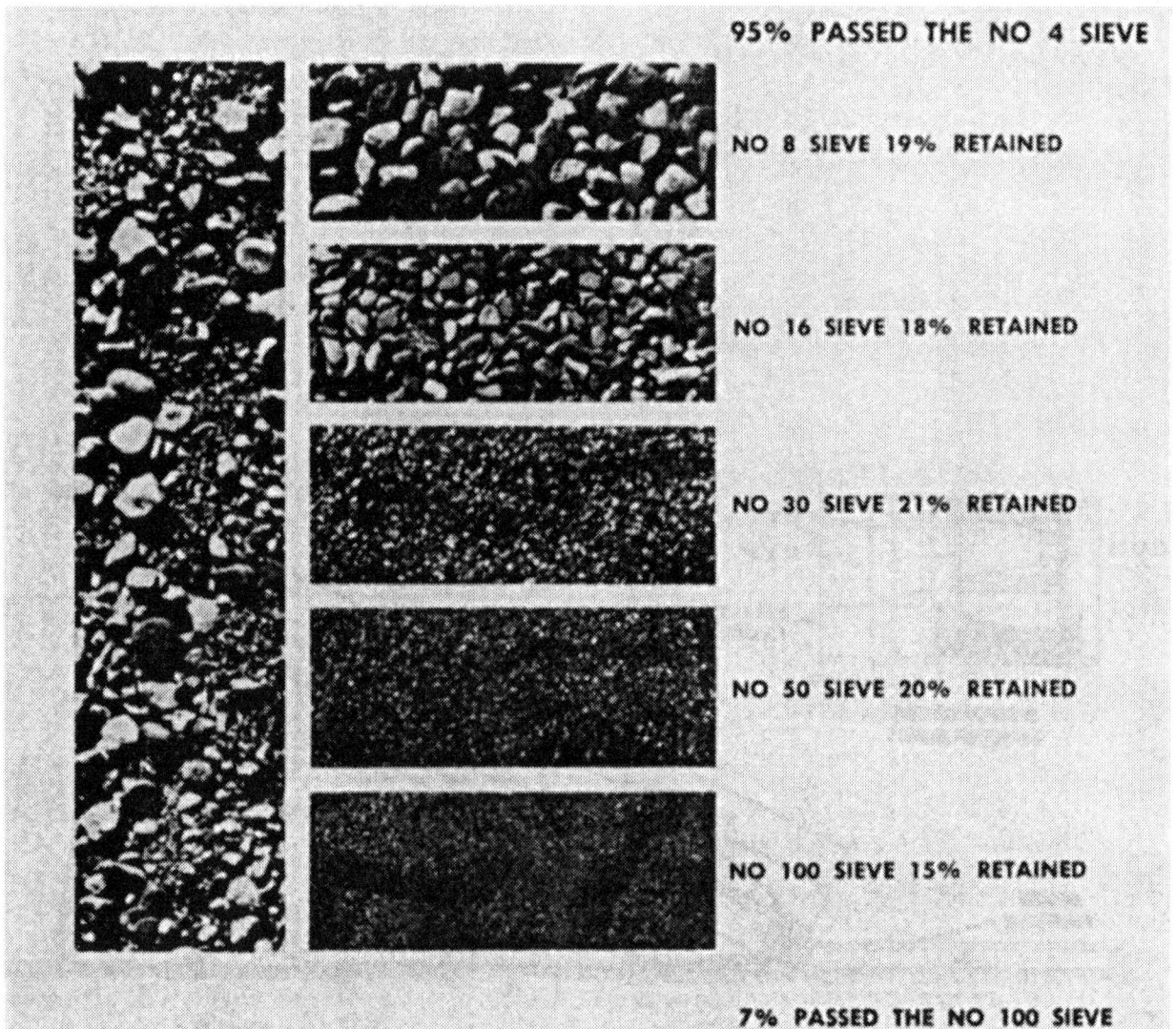

95% PASSED THE NO 4 SIEVE

NO 8 SIEVE 19% RETAINED

NO 16 SIEVE 18% RETAINED

NO 30 SIEVE 21% RETAINED

NO 50 SIEVE 20% RETAINED

NO 100 SIEVE 15% RETAINED

7% PASSED THE NO 100 SIEVE

*Figure 15. Well-graded sand.*

denser and stronger, bleeding is reduced, and finishing accomplished with greater ease. Because of these relationships a great deal of confusion is caused by the claims of manufacturers of secret formulas.

(c) Strive to resolve the use of admixtures to recognizable chemical elements, compounds, and substances; require manufacturers of proprietary materials to make known the basis of claims for their product.

(2) *Workability.* Considerably more water is required to make a workable mixture than is necessary for chemical reaction with the cement. Excess water reduces strength,

and causes segregation of materials and bleeding. The following additives are used to improve workability:

(a) Calcium or Ammonium Stearate—Up to 3 percent of cement by weight.

(b) Entrained air—3 percent to 6 percent of concrete by volume.

(c) Lime—Up to 6 pounds for each bag of cement.

(d) In addition to the above, saponifiers and oils are sometimes included in proprietary formulas.

(3) *Hastened setting.* A shorter setting time of concrete required to develop a usable strength may best be accomplished by use of high-early strength cement, at an

increase in cost of only 50 to 75 cents per cubic yard. Calcium chloride added to the mix at a rate of not more than 1 to 2 pounds per bag of cement will hasten the initial set and to a certain extent serve the same purpose. This material is one of the principal ingredients of several proprietary admixes. It is a salt; the use of quantities larger than the above amounts may be corrosive to the reinforcing steel. Calcium chloride should never be used in prestressed concrete.

(4) *Retarded setting.* The setting time of concrete may be retarded by use of cold water or cool aggregate; by shading the area or placing at night; and by use of commercial additives. Since the most favorable temperature for curing concrete is 70° to 80° F., there will be few instances when the above measures will not produce satisfactory results.

(5) *Densifiers.* Concrete may be made more dense and more waterproof by providing a well designed mix, decreasing water cement ratio, adding an extra bag of cement to the yard (about 7 bags per yard maximum), and/or 10 percent hydrated lime.

(6) *Durability.* To increase durability, air entrainment should not exceed 3 to 6 percent. This is accomplished by using air entrained cement or a commercial admixture. Air entrainment is a standard practice with industry. It is not intended to eliminate a well designed mix but to supplement it. Mineral surfacing materials are also used to increase durability of slabs and treads.

(7) *Reduced cost.* Fly-ash is used as a partial replacement for cement in a mixture to reduce the cost of concrete. On large dam projects it has been used also to reduce the heat of hydration, thus reducing the time required to cool mass concrete. Where strength is not important up to 35 percent of the cement in a mix may under certain conditions be replaced with fly-ash.

(8) *Proprietary admixes.* An open mind should be maintained regarding the use of commercial admixtures. These admixtures should not, however, be used except for specific purposes. Some manufacturers of these materials make extensive claims for their product, such as that it will increase workability up to 150 percent; increase strength up to 25 percent; reduce permeability 40 to 80 percent; increase bond to steel up to 40 percent; reduce shrinkage up to 20 percent; increase durability up to 400 percent; and the like. Since Government purchases are not made under trade names these materials can only be purchased by specifying required end results, or providing standard specifications for ingredients.

## 20. Concrete Mixtures

Select the proportions of materials for concrete mixtures to provide the quality of concrete needed for the intended use. Also use materials which will produce a plastic workable mixture capable of being handled and placed without objectionable segregation and without requiring excessive effort for satisfactory consolidation. Use appropriate admixture to provide specific qualities required. An excess of water weakens the concrete and tends to produce segregation. The ratio of water to cement within a normal mixture determines the strength of the concrete and has a profound effect on durability.

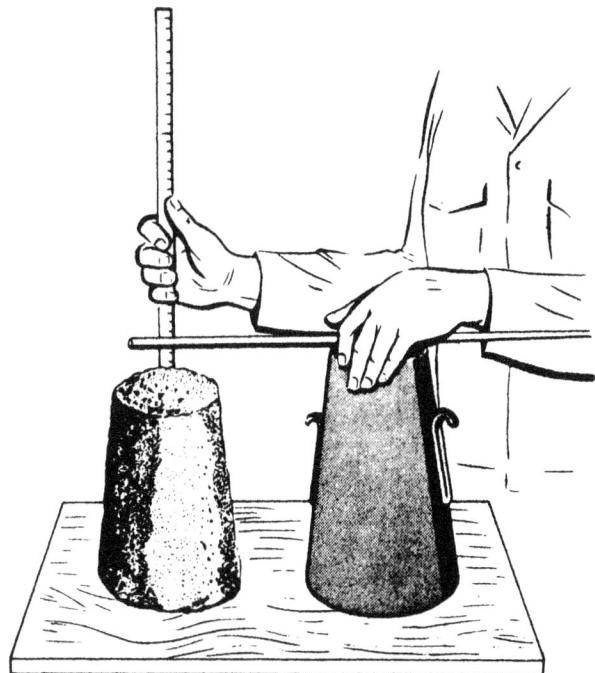

*Figure 16. Slump Test.*

The water-cement ratio for normal building construction should not exceed 7½ gallons per bag of cement and should reach this maximum only for heavy sections in mild climates. For thin sections in severe or moderate climates the water-cement ratio should not exceed 5½ gallons per bag of cement. After the water-cement ratio has been selected the proportions of aggregates and cement plus water should be adjusted to produce a mixture of proper consistency. The consistency, as measured by slump, (fig. 16) is shown in table I.

*Table 1. Slump Requirements*

| Type of construction | Slump in inches |
|---|---|
| General building construction | 2–3 |
| Thin, reinforced walls | 3–4 |
| Heavy-duty floor and slab construction | 1–2 |

Trial mixtures for various water-cement ratios based on dry, compact volumes of materials are:

## TRAIL MIXES

| Slump inches | Proportion of mix for 1-in. aggregate | Proportion of mix for 2-in. aggregate |
|---|---|---|
| | Water-cement ratio—5½ gal per bag | |
| ½–1 | 1:2:3 | 1:2:3½ |
| 3–4 | 1:1¾:2½ | 1:1¾:3 |
| 5–7 | 1:1½:2 | 1:1½:2½ |
| | Water-cement ratio—6 gal per bag | |
| ½–1 | 1:2¼:3¼ | 1:2½:4 |
| 3–4 | 1:2:3 | 1:2:3½ |
| 5–7 | 1:1¾:2½ | 1:1¾:3 |
| | Water-cement ratio—6¾ gal per bag | |
| ½–1 | 1:2½:3½ | 1:2½:4 |
| 3–4 | 1:2¼:3¼ | 1:2¼:3¾ |
| 5–7 | 1:2:3 | 1:2:3½ |
| | Water-cement ratio—7½ gal per bag | |
| ½–1 | 1:3:4 | 1:3:4¾ |
| 3–4 | 1:2½:3¾ | 1:2½:4¼ |
| 5–7 | 1:2¼:3½ | 1:2¼:3¾ |

Using coarse sand and one bag of cement the following are recommended weights of sand (fine aggregate) and gravel or stone (coarse aggregate). (This depends on the selected water-cement ratio, the maximum size of the aggregate and the particle shape):

| Water-cement ratio | Maximum size in. | Round | | Angular | |
|---|---|---|---|---|---|
| | | Sand | Gravel | Sand | Gravel |
| 5 gal/bag | ¾ | 185 | 230 | 185 | 185 |
| | 1 | 175 | 260 | 170 | 210 |
| | 1½ | 170 | 300 | 170 | 245 |
| | 2 | 170 | 340 | 170 | 280 |
| 6 gal/bag | ¾ | 240 | 270 | 235 | 220 |
| | 1 | 225 | 310 | 225 | 250 |
| | 1½ | 220 | 355 | 220 | 290 |
| | 2 | 220 | 410 | 220 | 335 |
| 7 gal/bag | ¾ | 300 | 310 | 295 | 250 |
| | 1 | 280 | 360 | 275 | 290 |
| | 1½ | 270 | 405 | 275 | 335 |
| | 2 | 270 | 465 | 280 | 380 |

*(Proportioned mix by weight / Pounds of dry aggregate)*

## 21. Mixing and Placing

*a. General.* Follow instructions in appendix II when mixing and placing a considerable volume of concrete. Machine mixing of concrete is preferable but hand mixing can be used extensively on small repair and maintenance jobs. Do not hand-mix concrete if the batch is over ½ cubic yard or contains more than can be placed and properly spaded into place in 30 minutes or less. Always do hand mixing on a tight wood platform, hard street pavement, or similar surface. A platform 10 feet square is large enough for a single batch mix.

(1) The following procedure is recommended for handmixing small batches.

(a) Measure desired amount of sand accurately and spread it 3 or 4 inches deep on the platform.

(b) Spread measured amount of cement over sand, and mix cement and sand until mixture has an even gray color; three complete turnings should be enough.

(c) Add water and mix until a smooth-working, even-gray mortar is obtained.

(d) Spread out mortar. Wet the coarse aggregate and add to mortar then turn the mass with hoes and shovels while adding small amounts of water; three or four complete turns should be enough. The resulting concrete should be a plastic, not sloppy, mess that slides, not runs, off the shovel and permits ready working in forms without objectionable separation.

(2) Another satisfactory hand-mixing method is to mix cement and sand, spread in a circle, add coarse aggregate and water and mix.

b. *Ready-Mixed Concrete.* When commercial ready-mixed concrete is to be employed, procure under ASTM Specifications C94. Mix each batch of concrete not less than 50 nor more than 100 revolutions at a mixing speed not less than 4 rpm after all materials are in the mixer drum. Neither speed nor capacity of mixer as used should exceed those recommended by the manufacturer. Do not add water to preserve workability. Procure concrete by class or strength. Make representative test cylinders for each project to assure the engineers that specified material has been furnished. Before awarding contracts for ready-mixed concrete, inspect the company's plant for: type of aggregate used, admixture used, method of controlling the mix, type and condition of mixers, and uniformity of strength as disclosed by local testing laboratories and large users of their product. Investigations have revealed that ready-mixed concrete may average about 15 percent less in strength than stationary mixed concrete. Two of the principal reasons are obsolete equipment and unreliable personnel on mixing trucks.

c. *Conveying.* Convey concrete from the mixer to the forms as rapidly as practicable using methods that will prevent segregation or waste of materials.

d. *Placing.* Work concrete into corners and angles of the forms and around the reinforcement

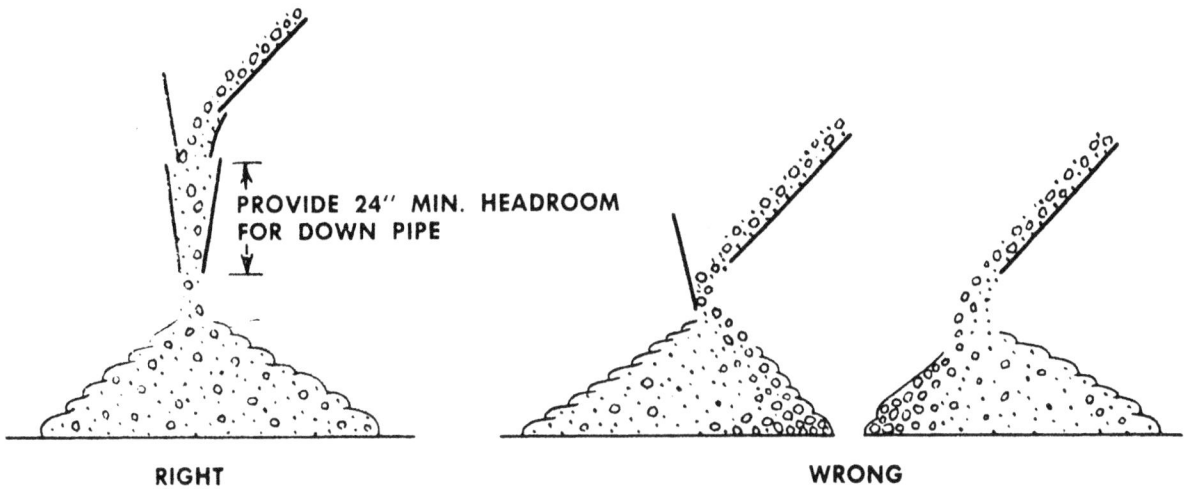

PROVIDE 24″ MIN. HEADROOM FOR DOWN PIPE

RIGHT

WRONG

1 Control of separation of concrete at end of chutes

*Figure 17. Placing concrete.*

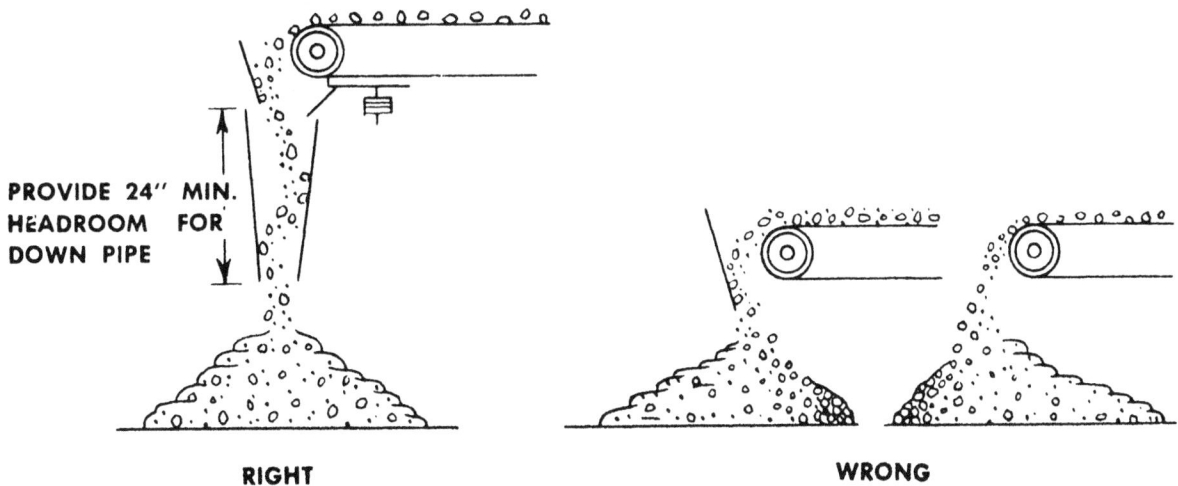

PROVIDE 24″ MIN. HEADROOM FOR DOWN PIPE

RIGHT

WRONG

2 Control of separation of concrete at end of conveyor belts

*Figure 17—Continued.*

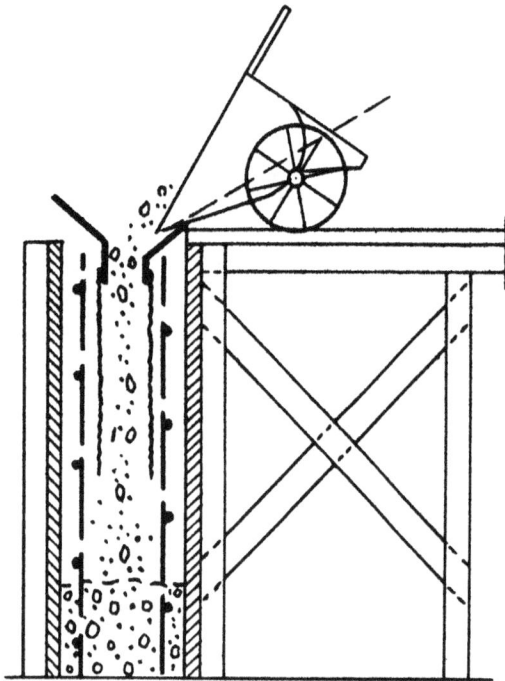

**RIGHT**

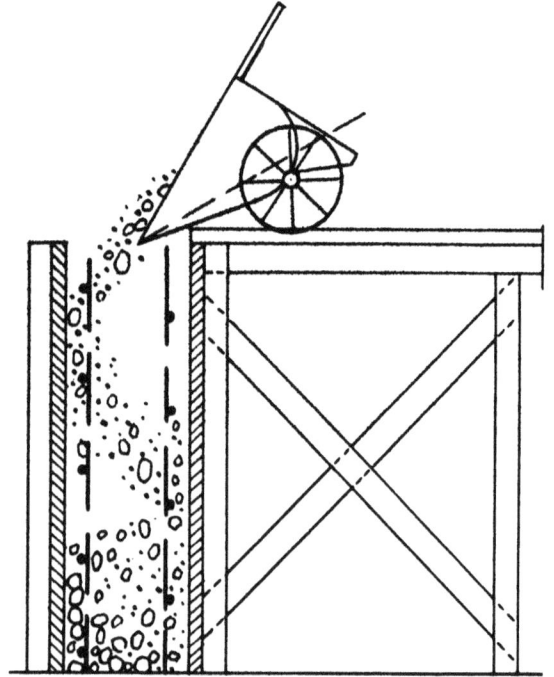

**WRONG**

3

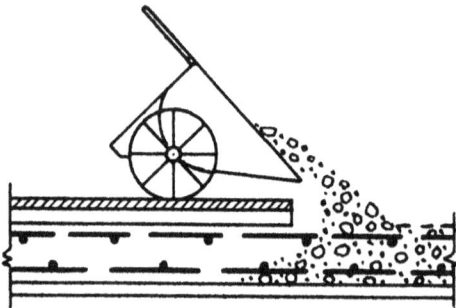

**RIGHT**

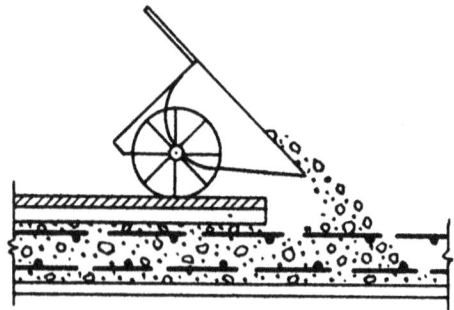

**WRONG**

4

3   Placing concrete in top of form

4   Placing concrete in slab

*Figure 17*—Continued.

AGO 10179A

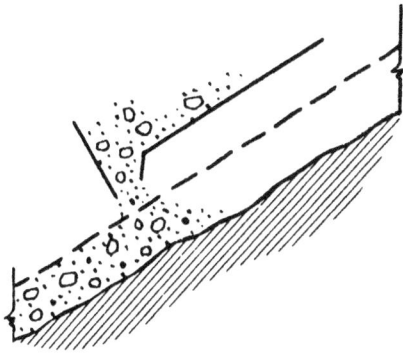

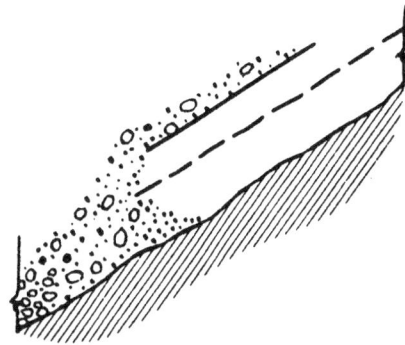

**RIGHT**                                        **WRONG**

5

5   Placing concrete on a sloping surface
*Figure 17—Continued.*

without causing segregation (fig. 17). Normally, not over 30 minutes should elapse between mixing and final placing, or not over 45 minutes from the time water is added until placed. Place concrete only on clean damp surfaces free from standing water, ice, dirt, or other foreign matter. Concrete should be consolidated by vibration or spading to achieve a continuous mass free from honeycomb and large voids of entrapped air. Vibration should not continue to the point at which segregation is induced as indicated by excessive water and mortar appearing at the surface. Placement should be carried on at such a rate that the formation of cold joints is prevented. If this cannot be done joints should be scarified and broomed with thin cement mortar prior to beginning a new pour.

*e. Under-Water Placement.* Concrete may be placed under water by use of a tremie (fig. 18); or by filling burlap bags with concrete and lowering into position. For either method low slump concrete is used. Avoid spading or vibrating concrete placed under water. Unnecessary motion tends to separate cement from aggregate. Compaction may be accomplished by use of a blunt-ended pole or rod 2″ to 3″ in diameter.

*f. Finishing.* Defective concrete (voids left by the removal of tie rods, rock pockets, ridges and local bulging on concrete surfaces permanently exposed to view, or exposed to water on the finished structure) should be repaired immediately after the removal of forms unless otherwise authorized or directed. Voids left by removal of the rods should be reamed and completely filled with low slump

mortar. Defective concrete should be repaired by cutting out the unsatisfactory material and placing new concrete, secured with keys, dovetails, or anchors. Do not permit excessive rubbing of formed surfaces. The finished specified should be provided on all exposed concrete.

## 22. Protection and Curing

*a. General.* Before placing, make sure all equipment and materials needed to protect and cure concrete

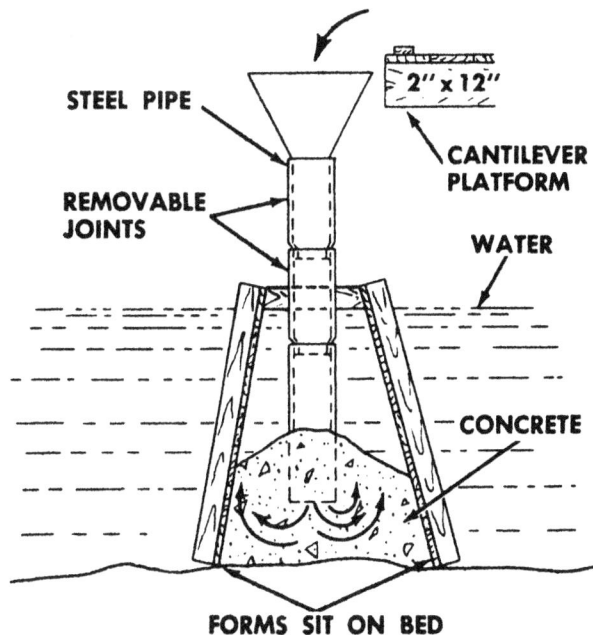

*Figure 18.   Placing concrete underwater with a tremie.*

are on hand. Take every precaution to protect fresh concrete from surface checking and cracking, excessive water loss, and damage by heavy rains, flowing water, mechanical injury, or extreme temperatures. Do not permit fires in direct contact with concrete at any time. Keep conduits and other openings through concrete closed during entire curing period to prevent circulation of air and possible checking. Cure concrete made with standard portland cement (types I or II) for at least five days, and concrete made with high-early-strength cement (type III) at least 3 days; curing twice as long may be necessary in extremely low or high temperatures. Depending on availability of materials and local conditions, protect and cure concrete as soon as it has set sufficiently to prevent damage. Use one or more of the methods below.

*b. Water Curing.* Keep concrete continuously wet with water-saturated material, mechanical sprinklers, flooding, or other means. Do not permit periodic wetting with alternate dry periods. If wood forms are left in place during curing, keep them wet to prevent joints from opening and concrete from drying. Use clean water, free from any substance which might cause objectionable staining or discoloration.

*c. Saturated-Sand Curing.* Cover concrete surface with at least one inch of sand. Keep sand spread evenly and saturated continuously during curing period.

*d. Fabric Curing.* Cotton quilts for curing concrete are made of two sheets of heavy, coarse-woven fabric stitched over a layer of low-grade cotton batting. Lay cotton quilts with lapped joints; keep them wet.

*e. Paper Curing.* Kraft paper for curing concrete is usually two-ply, reinforced between plies with fibre strands in two directions and sealed with asphalt. Lay kraft sheets with 4-inch laps. Cement joints with waterproof glue or hold them together with heavy gummed-paper strips.

*f. Curing Compounds.* Apply curing compounds in accordance with applicable specifications listed in appendixes I and II. Do not use curing compounds on surfaces to which additional concrete is to be bonded or on vertical surfaces exposed to sunshine and wind. In cold weather, curing compounds should not be used on concrete surfaces which are kept at satisfactory curing temperatures by steam pipes or other means.

## 23. Cold Weather Concreting

*a. General.* When concrete work is done in freezing weather, pay special attention to conditioning aggregates, mixing and placing, preparation of forms and subgrade, and protection of concrete during and immediately after the setting-up period. When outside temperature falls to 32°F., expansion of water in unprotected concrete exerts enough force to destroy the bond between cement and aggregate, making the concrete worthless (fig. 19). Make every effort to mix and place concrete during periods when the temperature is at least 40° F. and rising.

*b. Storage and Protection of Materials.* Keep concrete materials intended for use at freezing temperatures in closed buildings or under sheds, covered with tarpaulins or otherwise protected to prevent accumulation of moisture and ice. Use hot mixing water where practicable. If aggregates are stock-piled in the open, heat both sand and coarse aggregate enough to thaw out frost and ice before placing them in the mixture. One method of heating aggregates is:

(1) On a site convenient to aggregate stock pile, lay a length or two of corrugated metal or reinforced concrete culvert pipe with one end toward prevailing winds to give a good draft, (fig. 20).

(2) Build a fire inside pipe.

(3) Cover heated pipe with aggregates, sand at one end and coarse aggregates at the other.

(4) Shovel cold aggregates to pipe and over it, always removing heated aggregates from opposite side.

> *Note.* Take care to keep the fire from spreading to adjacent buildings, materials, or equipment.

*c. Cold-Weather Protection of Above-Ground Concrete.* Protect above-ground concrete mixed and placed in freezing weather with temporary inclosures, such as, fire-retardant tarpaulins heated with warm air or steam. Maintain a temperature of at least 50° F. for at least 5 days when standard portland cement is used and 3 days when high-early-strength cement is used. Longer protection periods are better if construction schedules permit. Do not stop cold-weather protection in a manner which will subject concrete surfaces to sudden temperature drop of more than 25° F. Determine temperature of concrete surfaces and surrounding air by hanging thermometers outside the concrete.

*d. Concrete on Ground.* Protect concrete slabs placed on the ground in freezing weather by keeping them covered with straw, dirt, tarpaulins, or other materials. Continue the protection for the same length of time as with above-ground concrete.

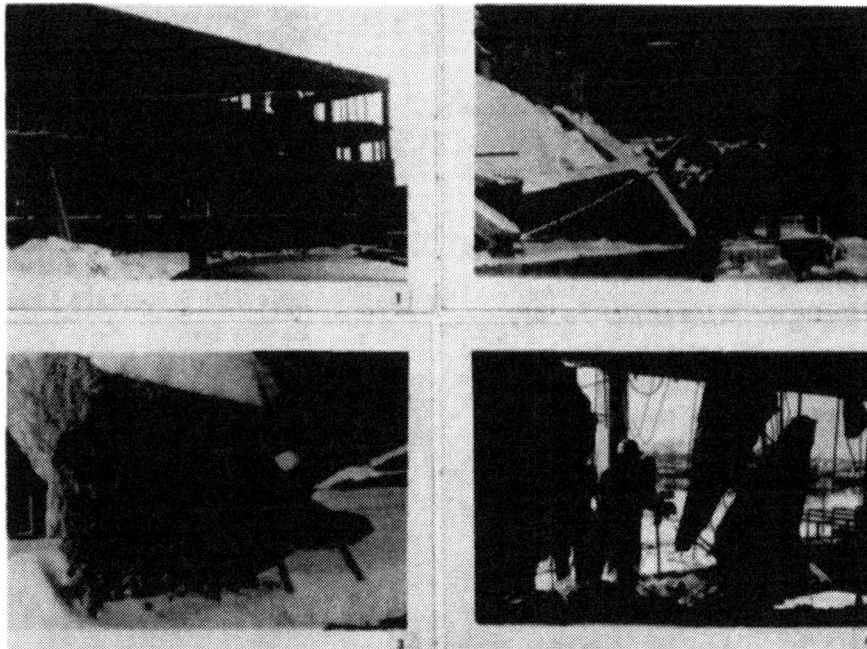

1 Building looked fine until thaw came; shoring was removed about 4 weeks after concrete was placed

2 Failure did not occur until warm rains came, about 6 weeks after placing

3 Note short dowels at base of column that failed when roof collapsed

4 Removal of broken slab. This is a dangerous job which must be done carefully to avoid further collapse of structure

*Figure 19. Failure of frozen concrete. (Freezing of concrete before settling caused serious damage to a three-story building in which columns, floor-slab, and roof were of reinforced concrete.)*

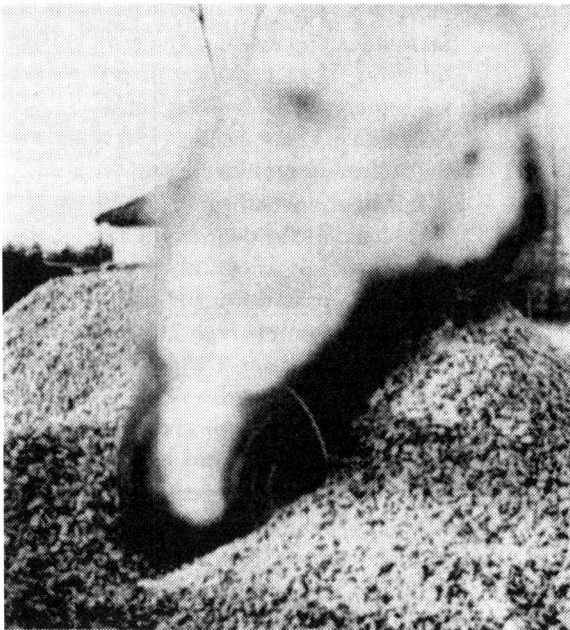

*Figure 20. Heating aggregate.*

e. *Frozen Subgrade.* Do not place concrete on a frozen subgrade as uneven settlement resulting from thawing of the ground may cause through-wall cracks. Protect all dirt subgrades over which concrete is to be placed by coverings of straw, hay, tarpaulins, cotton mats, or other material. Place coverings immediately after completing subgrade excavation. Do not remove them until just before concrete is to be poured.

f. *Removing Frost and Ice.* Shortly before placing concrete, remove all frost and ice coatings from reinforcing steel and inside surfaces of forms. Steam or hot water jets are recommended for this purpose. Observe safety requirements when using steam jets.

g. *Admixtures to Lower Freezing Point.* Avoid the use of salts, chemicals, or other foreign materials to make concrete set faster or to lower its freezing point. The quantity of material required to lower the freezing point appreciably is so great that the desirable qualities of the concrete will be adversely affected. High-early-strength cement or calcium chloride in an amount not to exceed 2 percent by weight of the cement may be used when there is a demand for early usage of concrete.

## 24. Concrete Work in Hot Weather

Sunlight, wind, and high temperatures affect concrete mixed and placed in hot weather. When concrete is to be placed during extremely high temperatures take the following precautions:

*a.* Cool aggregates stored in the open by spraying them frequently with water.

*b.* Use coolest water available at site for mixing.

*c.* During extremely high temperatures, schedule work to allow time for one unit to begin to cool before placing the next adjacent unit.

## 25. Special Classes of Concrete

*a. General.* There are many types and classes of concrete. All classes, including Classes D, E, and F, Lightweight Concretes, are covered thoroughly in appendix II. Only "Porous Concrete," "Concrete Fills," and "Pneumatically Applied Concrete," will be discussed here.

*b. Porous Concrete.* Porous concrete is used mainly for drain tile and drainage bases under lined canals, spillways, and slabs, or other work on grade where a relatively free passage of water is necessary.

(1) *Base courses.*

(a) Make base courses for porous concrete using the same aggregate as for standard concrete. Aggregates should pass a ¾-inch square mesh sieve and be retained on a ³⁄₁₆-inch square mesh sieve. Approximate proportions are one part portland cement to five parts aggregate by weight. Use only enough water so the cement paste thoroughly coats and binds aggregate parts but does not fill voids in the aggregate. Proportions may be varied when necessary, provided that the finished product meets requirements for compressive strength and porosity ((3) below).

(b) Mix carefully to prevent cement from lumping and accumulating on mixer blades and sides. Avoid excessive tamping; it over consolidates the mass, making it less porous.

(2) *Drain tile.*

Prepare porous concrete for drain tile in the same way as for base courses except that all aggregate should pass a sieve with ⅜-inch square mesh openings.

(3) *Tests.* Make, store, and test 6- by 12-inch test cylinders in accordance with American Society for Testing and Materials (ASTM) designation C 39, Compressive Strength of Molded Concrete Cylinders. Compressive strength at 7 days should be at least 1,000 pounds per square inch. Porosity of concrete at 7 days should permit water to pass through a 12-inch slab at not less than 7 gallons a minute for each square foot of slab, with a constant 4-inch depth of water on the slab.

*c. Concrete Fills.* Concrete fills can be used to advantage on concrete roof slabs and in the spaces between wood sleepers when wood floors are laid over concrete slabs above grade.

(1) *Sawdust-concrete roof fill.*

(a) When a lightweight insulating concrete is desired for a roof fill, sawdust concrete may be used.

(b) To make good nailing concrete, mix equal parts by volume of portland cement, sand, and pine sawdust with enough water to cause a slump of not more than 2 inches. If the resulting concrete is too hard for proper nailing, increase the sawdust content up to 50 percent. Mix concrete thoroughly, moist-cure it in place for 2 days, and allow it to dry for at least 1 day before nailing. Cinders, asbestos fiber, and various patented materials, such as expanded mica and foaming compounds, can be substituted for sawdust in making concrete suitable for receiving and holding nails.

(c) Nailing concrete 28 days old must meet the following requirements:

*1.* Penetrative nailing property equal to that of short-leaf pine.

*2.* At least 185 pounds average withdrawal resistance for five tenpenny common nails driven 2½ inches and at least 165 pounds withdrawal resistance for any single nail.

(2) *Lightweight insulating concrete.* Lightweight insulating concrete should be of a type that can be readily placed, finished, and cured. Install expansion or ventilating joints at least 1 inch thick through the full depth of the fill, at the junction of the fill with fixed vertical surfaces and at intermediate intervals not exceeding 50 feet in each direction. Provide an edge venting system capable of

intercepting and releasing internal vapor pressure from the joints without exerting internal pressure against built-up roofing, where such is to be installed. Fill joints with a low-density glass-fiber or mineral-wool insulating board conforming to Federal Specification HH-I-526 or HH-I-562 to permit free passage of moisture vapor. The lightweight insulating concrete should consist of type I, type II or type III portland cement, water, and either light-weight aggregate conforming to the ASTM Specifications C 332 or preformed chemical form. The unit weight of the concrete when 28 days old should not exceed 60 lbs. per cu. ft. when in the oven-dry condition. The compressive strength should be at least 125 psi, and the material should be capable of supporting a load of 140 lbs. applied to a circular area $1\frac{1}{8}$ in. in diameter without evidence of crushing.

(3) *Tar-concrete floor fill.* Always use tar-concrete fill on slabs laid on the earth subgrade. It may also be used on slabs constructed above grade. It retards rising moisture and possible dry rot in wood sleepers and flooring.

  (a) *Materials.* When making tar-concrete fill, use 40 to 50 gallons of tar, Federal Specifications R-T-143, to each cubic yard of sand. Before tar is added, heat sand enough to dry it thoroughly. Heat tar before mixing in sand so temperature of the mass will not exceed 225° F. when sand is added. Thick yellow smoke rising from tar indicates overheating.

  (b) *Installation.* To install the fill—

    *1.* Set sleepers (pressure treated to prevent decay) and check with builder's level. Leave sufficient space between rough slab and sleeper bottom for wedges to hold sleepers level. Anchor sleepers to slab with clips or wire anchors.

    *2.* Place tar-concrete fill between sleepers, forcing material under them to provide a firm bearing. Fill flush with top of sleepers. If floor is laid over slabs resting on earth subgrade, place a layer of 15-pound asphalt, or tar-

saturated felt between tar-concrete fill and finished flooring.

  (c) *Safety.* Tar-saturated mops, brooms, and rags, are fire hazards. Store them in metal containers or place them in the open some distance from buildings and inflammable materials. Observe all safety requirements when heating, mixing, and placing hot mixtures.

(4) *Cement-concrete floor fill.* Use cement-concrete fill only on concrete slabs constructed above grade.

  (a) *Materials.* Use one part portland cement and four parts sand or other suitable fine aggregate for cement-concrete fills.

  (b) *Installation.* Before placing the fill, install floor sleepers as outlined above. Pour cement-concrete fill between sleepers, forcing mortar under them to provide full bearing. Mortar consistency must permit flow under sleepers. Screed the fill level with sleepers; take care that finished fill is no higher than sleepers at any point. Place asphalt-saturated membrane paper or polyethylene sheet between concrete fill and finished flooring.

  (c) *Pneumatically applied concrete.* Mortar and concrete containing aggregate up to $\frac{1}{4}$ inch in size can be shot into place, as floor fill, using air pressure. The aggregate used for such work should normally be damp (approximately 5 percent free moisture) so as to pick up the cement. Recommended practice for pneumatic application ("Shotcrete") is contained in the U.S. Bureau of Reclamation Concrete Manual, the ACI Manual of Concrete Inspection, and Portland Cement Association publications No. ST-17-2, and CB-3.

## 26. Painting Concrete

When required, paint concrete in accordance with TM 5-618 and paragraph 61 of this manual. Good dense concrete with adequate cement content, and scrubbed-in cement-water-grout filler (Federal Specifications TT-P-21) properly cured, will very seldom need paint coatings except for utilitarian and decorative purposes.

## 27. Precast Concrete Piles

*a. General.* Technical Manuals 5-622 and 5-624 provide guidance for the construction, handling, and driving of concrete piles.

*b. Maintenance.* Little maintenance is required on concrete piles. When required, it should be accomplished as outlined in Section VII, Maintenance and Waterproofing of Concrete Walls.

## 28. Other Precast Concrete Items

In addition to precast concrete piles, precast floor and ceiling beams, lintels, joists, sills, roof slabs, steps, tanks, manholes, and similar items, may be obtained for use in maintenance work (fig. 21). These items should be certified as having been constructed of concrete of quality appropriate for the intended use. The procurement of such items may include those that have been prestressed. Prestressed concrete is that in which the reinforcing steel is specially stressed to induce desired compressive stresses in the hardened concrete. Prestressing permits more efficient use of high-strength steel and concrete. It also prevents cracking due to shrinkage of the concrete. In pretensioned, prestressed con-

1   Mason laying solid top block for bearing course to support precast concrete joists

*Figure 21.  Precast concrete joist construction.*

crete the steel is stretched before the concrete is placed. When the concrete has hardened sufficiently, the applied tension is released and the steel is held in tension through bond with the concrete. This method is the one normally used for precast members. Prestressed, precast items should be the product of plants having facilities for careful quality control of the concrete batching, mixing, placing, and curing and for installation, tensioning, and protection of prestressing tendons. Prestressing steel should have at least 1½ in. of concrete cover. Precast concrete items shall have an absorption of not more than 8 percent by weight after immersion in water for 48 hours. Care should be taken to insure that the precast items are not overstressed during transportation and installation.

2  Precast concrete joists well identified in accordance with joist layout plans

*Figure 21*—Continued.

3   Paper-backed welded wire fabric is sometimes used as combined forming and reinforcing for the slab

*Figure 21*—Continued.

4   Wood sheathing being placed over form supports
*Figure 21*—Continued.

5   Section of completed floor
*Figure 21*—Continued.

# SECTION VI

# CONCRETE SLABS AND FLOORS

## 29. Slabs

*a. General.* Concrete slabs on grade are so integrated with foundation walls, piers, columns, and footings that they become a part of the foundation. They transmit dead and live loads to the soil. Slabs are poured in place on bases prepared as described in paragraph 8.

*b. Requirements.* Use plain, unreinforced flat slabs to carry uniformly distributed light loads, usually not over 100 psf. Welded-wire mesh-reinforced slabs are used for loads up to 500 psf. Slabs with bar reinforcement are used for heavier or concentrated loads. Mixing, placing, finishing, curing, and surface treatment of slabs on grade are similar to those for concrete floors over wood joists construction (par. 30). (For heavy duty floor slabs see par. 31.) Prepare and construct exterior slabs for roads, runways, and miscellaneous pavements in accordance with instructions in TM 5–624.

*c. Expansion Joints.* Provide expansion joints between new slabs on earth and existing slabs on earth. Also, between slabs on earth and all vertical surfaces; provide additional expansion joints as actually required by the thermal expansion of the concrete over temperature range expected. Except for insulated joints the joint material should be premolded expansion joint filler strips. Normally, joints should be ½ inch thick and the full depth of the slab. Construction joints are generally placed at expansion joints.

*d. Contraction or Control Joints.* Depending upon the location of expansion joints, contraction or control joints should be provided to control cracking by pouring slabs in alternate checkerboard sections approximately 800 square feet in area. An alternate method is to place the slab continuously, as limited by expansion and construction joints, and to saw the required control joints with an approved concrete saw. Sawed joints should be cut ⅛-inch wide and approximately ¼ of the slab thickness in depth. Remove saw cuttings and debris from the joint. Insert fiber board strips ⅛-inch thick, of the required width and length. Omit crack-control joints perpendicular to radial-grade-line joints in floor slabs in theatre auditoriums. Provide radial-grade-line joints to conform with seat installation layout. When crack-control joints are omitted, use

care in selection of mixture proportions, in control, inspection of placing, and curing concrete slabs on grade.

## 30. Concrete Floor Over Wood-Joist Construction

*a. General.* Wide cracks, uneven surfaces, excessive splintering, and other faults in flat-grain softwood floors cause unsanitary or hazardous conditions. Applying a concrete-finish wearing surface on the floor corrects these problems. These two different materials can be successfully combined to act together structurally, the concrete resisting the flexural compressive stresses, and the wood the flexural tensile stresses. When so combined the two greatest forces that tend to produce failure are the shearing action at the plane between the two materials, and the vertical force which tends to cause separation and buckling at the same plane. Nails partly driven in wood floors are successful in resisting both of the forces. This section describes the materials, equipment, and labor needed to install a concrete-finish floor 1½ inches thick over a faulty wood floor on wood joists (fig. 22). Follow instructions closely for best results. Assure that floor framing is of such design, structural strength, and rigidity to withstand, without appreciable deflection or movement, the maximum load normally imposed. Floors for ordinary usage are designed for only about 40 lbs. per square foot (psf) live load. Since 1½″ thick concrete will add 20 lbs. psf additional temporary or permanent underfloor shoring may be required. After concrete is completely cured a considerable increase in floor rigidity and a slight increase in total floor strength may be gained. The addition of temporary floor girders, while the concrete is fluid, will help prevent normal joist deflection. Control the mixing, placing, troweling, and curing operations carefully to insure a surface which withstands hard wear and is free from excessive dusting, pitting, crazing, or other defects. Pay particular attention to proper proportioning of materials, use of minimum amounts of mixing water, minimum wood-floating, and proper timing of final steel-troweling. Observe all instructions in this section because defective surfaces can rarely be brought up to standard after concrete has set.

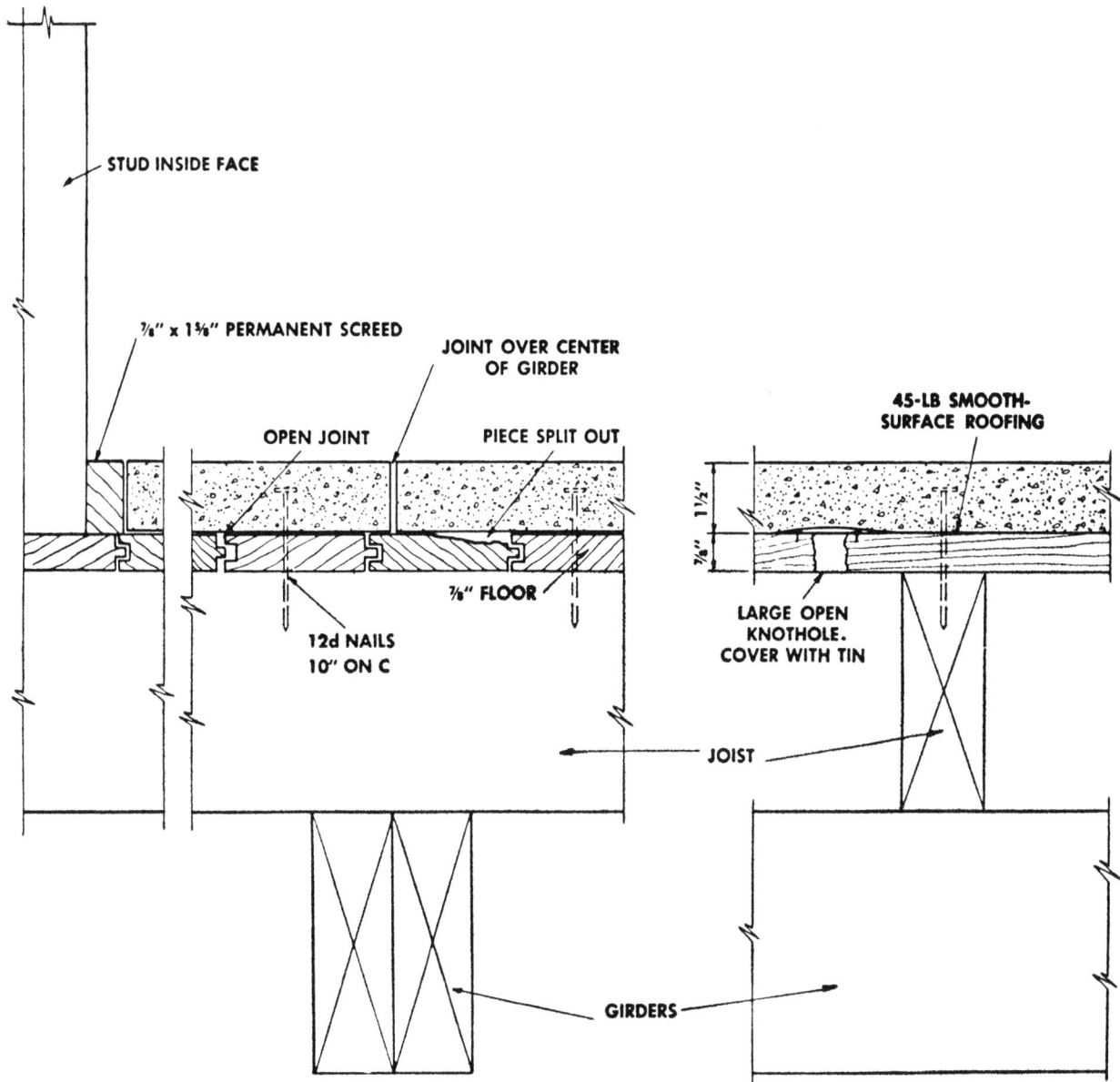

STUD INSIDE FACE

⅞" x 1⅜" PERMANENT SCREED

JOINT OVER CENTER OF GIRDER

OPEN JOINT

PIECE SPLIT OUT

45-LB SMOOTH-SURFACE ROOFING

⅞" FLOOR

12d NAILS 10" ON C

1½"

⅞"

LARGE OPEN KNOTHOLE. COVER WITH TIN

JOIST

GIRDERS

*Figure 22. Concrete-finish floor laid over rough flat-grain wood floor.*

*b. Preparatory Work.*

(1) *Removing equipment and raising fixtures.* Where practicable move all equipment before beginning floor-correction work, so that it will not interfere with the work. Make any changes in levels of floor drains, and other changes in plumbing, heating, and other utilities required by the change in floor level. Separate all new and existing pipes that will pass through the floor from the flooring by wrapping with at least two turns (plies) of asphalt saturated felt. Leave sufficient clearance to permit free movement. Trim the insulating felt flush with the finished concrete flooring. Retard the passage of air by loose packing the space between the flooring and pipe with glass-wool, rock-wool, or other fire-resistant material. Give metal surfaces in contact with the flooring a coating of bituminous paint.

(2) *Repairing cracks and holes.* Nail all loose ends and springy boards securely. Place sheet metal or other material over all open knotholes and cracks which are wide enough to leak mortar.

(3) *Cleaning.* Clean all dust, dirt, loose particles, and debris from wood which is to receive the concrete wearing surface. Remove oil and grease spots by the process described in paragraph 35c(1).

(4) *Soaking wood floor.* Thoroughly soak wood floors with water 24 hours before pouring concrete. If building floors are to be washed frequently, restrict water from passing through the concrete surfacing to wood underflooring by covering wood with a 45-pound smooth-surface roofing material laid across floor boards with coated side up. Extend roofing material up intersecting walls about 4 inches. Lap edges 4 inches and secure them with cold-lap asphalt cement. Do not water-soak flooring on which roofing material is laid.

(5) *Reinforcing.*

　(a) *Nail reinforcing.* Nail reinforcing, as described below, may be used as an alternative to wire mesh or metal lath. Wire mesh is preferred to metal lath for areas greater than 100 square feet. Drive twelvepenny (3¼-inch) or longer common wire nails at 10-inch intervals along center lines of all joists. Leave a full

inch of nail extending above upper surface of wood floor. A 1-inch board held along the nail can be used as a combination straight edge and gage (fig. 22).

　(b) *Mesh reinforcing.* Lay mesh reinforcing with lapped and wired edges over the wooden floor or roofing surface. Secure the reinforcing by nails or staples placed about 6 inches apart in two directions. Drive the fastenings so that the reinforcement is not embedded in the roofing and so a gage $\frac{1}{16}$ inch thick will fit between the roofing and the reinforcing. Extend nails through not less than $\frac{7}{8}$ of the wood floor thickness and, in case of a double floor, into the lower layer. Use 6 x 6/8 x 8 gage wire mesh. Use expanded metal lath, regular nonflattened industrial mesh, diamond pattern, size approximately $\frac{3}{4}$ inch x $1\frac{1}{2}$ inch, weight about $\frac{3}{4}$ pounds per square foot.

(6) *Flashing.* Provide adequate perimeter flashing or concrete cove to prevent leakage to the space below in areas subjected to water spillage.

(7) *Screeds.* To control area and depth of concrete finish, nail to the floor $1\frac{5}{8}$ x $\frac{7}{8}$-inch wood screed strips (actual dimension) surfaced on four sides. Make sure strips are free from large loose knots or other surface imperfections.

　(a) *Permanent screeds.* Nail permanent screeds against studding around entire perimeter of area to be floored.

　(b) *Intermediate screeds.* Nail necessary intermediate screeds to form panels controlling the pour and to form construction joints in the finished concrete floor. Form all construction joints directly over centers of supporting girders.

*c. Materials.*

(1) *Cement.*

　(a) Use portland cement conforming to Federal Specifications SS–C–192a, type I or type II, for floors where normal traffic can be kept off at least 7 to 14 days depending on type of usage.

　(b) Use high-early-strength portland cement conforming to Federal Specifications SS–C–192a, type III, for floors where normal traffic must be resumed after 3 days.

**Caution: Use only one brand of cement on any one job.**

(2) *Lime.* Use hydrated lime for structural purposes conforming to Federal Specification SS-L-351.

(3) *Aggregates.*

(a) *Fine.* Use sand conforming to Federal Specification SS-A-281, aggregate; for portland-cement concrete, class 1.

(b) *Coarse.* Use stone conforming to Federal Specification SS-A-281b aggregate; for portland-cement concrete, class 2, except for grading. This material is similar to coarse aggregate used in the seal coat for bituminous surfaced roads. Use the following grades:

| U.S. standard sieve size | Percent by weight passing each sieve |
|---|---|
| ½ inch | 100 |
| ⅜ inch | 85 to 100 |
| No. 4 | 10 to 30 |
| No. 8 | 0 to 10 |

(4) *Water.* Use potable water for mixing.

d. *Proportions and Mixing.*

(1) *Cement and aggregates.* When materials are measured loose, use the following proportions by volume:

| | Cement | Fine aggregates | Coarse aggregates |
|---|---|---|---|
| Dry materials | 1 | 1⅓ | 2 |
| Materials when moisture is over 3 per cent by weight. | 1 | 1½ | 2 |

If the sand contains a greater proportion of fines than the specifications call for, use one-sixth part less sand by volume than shown above. If coarse aggregate is more or less uniform in size, reduce quantity to not less than 1½ parts.

(2) *Lime.* Use 6 pounds of hydrated lime for each bag (94 pounds) of cement. Lime acts as a waterproofing agent and also makes concrete more workable.

(3) *Water.* Do not use more than 5½ gallons of water, including water in the aggregate, for each bag of cement. The mixture as taken from the mixer should be plastic but without free surface water and should have a slump of 2 or 3 inches.

(4) *Mixing.* Mix materials 2 or 3 minutes. After each mixture, empty mixer completely and free all lumped or caked materials adhering to mixer drum or arms.

e. *Placing.* Place concrete surface material as follows:

(1) Dump material from mixer into wheelbarrows or other containers from which it is to be applied, or into a watertight box from which it can be shoveled into buckets or wheelbarrows. Avoid separating materials during rehandling. Empty box completely after every third batch.

(2) Drop each batch of concrete as near as practicable to point of final finish. Deposit concrete over surface to be covered, beginning at point farthest from building entrance.

(3) Place each panel in one continuous operation without interruption.

(4) As each panel is filled, strike its surface flush and consolidate the concrete by floating. Avoid bringing water to surface.

f. *Finishing.*

(1) *Plain finish coat.*

(a) *Materials.* To prepare a plain finish coat, use sand and cement described in c above.

(b) *Mixing.* Use one part cement to two parts fine sifted sand. Mix dry materials thoroughly before applying them.

(c) *Application.*

1. Immediately after floating concrete, spread dry-finish mixture uniformly over surface of slab, ½ pound for each square foot of floor area. Work coating into surface with a wood float. Do no additional dusting.

2. Allow slab to become reasonably hard before steel-troweling to final surface. To provide compaction, apply pressure to trowel, as it moves over the surface. The longer final troweling is delayed the better the resulting surface, provided the concrete does not set before troweling is completed. Take care not to delay steel troweling so long that entire slab cannot be covered before final set occurs. A thin slab will set faster than a thick slab (fig. 23).

(d) *Advantages and disadvantages.* This type of finish permits a reduction in the finishing time when high slump concrete is inadvertently received. Also, it requires slightly less skilled finishers, especially for lean mixes. If too much cement is used a checking or crazing of the surface develops; therefore, a monolithic finish should ordinarily be used except where floor covering is to be installed.

(2) *Monolithic finish.* Tamp concrete with a suitable tool to force aggregate down from the surface. Then screed with a straight edge and rough float to the required finish level. While concrete is still green, but sufficiently hardened to bear a man's weight without deep imprint, float it so that the coarse aggregate is invisible. Use sufficient pressure on floats to bring the moisture to the surface. Trowel the concrete until smooth, slick, dense and free of tool marks (fig. 23). This method is the most desirable procedure and is almost universally used, but it requires concrete of optimum type.

1

2

1  Steel trowler and edger      2  Troweling operation

*Figure 23. Steel finishing tools and troweling operation.*

(3) *Colored finish.*

**Caution: Repairs to floors with such finish are very difficult.**

(a) *Material.* Use the following materials for a colored finish coat:

1. *Cement.* Use cement described in paragraph 19a.
2. *Sand.* Use well-graded sand with not less than 80 percent passing a No. 8 sieve and not more than 10 percent passing a No. 50 sieve.
3. *Pigment.* Commercially pure red iron-oxide pigment is generally used because it is usually available and comparatively low in cost. Other colors may be used, giving first consideration to availability and cost.

(b) *Mix.* Use 1 part cement and 1½ parts sand with 10 pounds of iron oxide for each bag of cement. Mix dry materials thoroughly before spreading.

(c) *Application.*

1. Spread coating mixture evenly over floor slab immediately after wood floating is completed. Use at least 100 pounds of pigmented mixture per 100 square feet of floor area. After spreading mixture, float it lightly until moisture begins to appear on the surface.
2. Do final steel-troweling as in *f(1)(c)* above.

(d) *Integral colored topping.* Finish as described in *f(1)(c)* above, but omit steel troweling. Mix cement and sand in proportion of 1:2, adding a minimum amount of coloring to give desired color. Sand usually sold as "plaster sand" or "masonry sand" gives good results because of fine gradation. Mixture should be of thick cream consistency. Spread sufficient mixture over floor to provide ⅜ inch to ¾ inch thick topping and finish to smooth slick finish. This method requires more skill and is more expensive, but provides a longer wearing colored surface than the dusted treatment. Too much coloring will weaken the concrete. Follow manufacturer's instructions for the pigments used. If too much time expires between placing base concrete and colored topping there will sometimes be a separation of material when hardened. Following is a list, with appropriate specification numbers, of the six most popular colors used as pigments for colored concrete floor finishes:

| Shades of color | Pigment designation | Specifications | |
|---|---|---|---|
| | | ASTM | Federal |
| Grays to Black | Black oxide of Iron or Carbon Black. | D-769 D-561 | TT-I-698 |
| Blue_ _ _ _ _ _ _ _ | Ultramarine Blue_ _ _ _ | D-262 | TT-U-450 |
| Bright Reds to Deep Red | Red Oxide of Iron_ _ _ | D-84 | TT-I-511a |
| Brown, Ivory, Buff or Cream | Brown Oxide of Iron_ Yellow Oxide of Iron. | D-768 | TT-I-702 TT-Y-216 |
| Green_ _ _ _ _ _ _ _ | Chrome Oxide of Greenish Blue Ultramarine. | D-263 D-262 | TT-C-306 TT-U-450 |

(4) *Other finishes.* Concrete may be finished to best meet the need for which placed. Generally inside surfaces should be finished as smooth and dense as possible, and outside surfaces given a broom finish to prevent slippage during wet and icy weather. Treads and platforms should be sprinkled with 1/4 pound per square foot of fine emery dust or other abrasives prior to final brooming or troweling. Prepared epoxy-bound skid-resistive finish material for existing concrete surfaces may be purchased under specification MIL-D-23003. Power-driven finishing machines may be used for large areas.

*g. Hardeners.* Integral floor hardeners can be used to prevent dusting where curing cannot be properly controlled or where sand is known to be deficient in fine aggregates (par. 36). Calcium chloride or a suitable type of commercial product is recommended. Dry-flake calcium chloride can be used at the rate of 2 pounds for each bag of cement. Place the calcium chloride in the mixer-skip with the aggregates. Be careful in using calcium chloride in hot weather, especially with high-early-strength cement because too rapid a set may occur.

*h. Control Joints.* When making control joints, extend surface scoring to a minimum depth of 1/2 inch. Be sure that scoring along construction joints is directly over supporting girders and in line with them.

*i. Protecting Walls.* Use an apron skirting of building paper to prevent staining or spotting finished wall surfaces, and avoid structural damage to wall surfaces.

*j. Curing.* As soon as water can be sprayed over the concrete surface without pitting or roughening it, spray the slab lightly. Keep surface damp for 24 to 48 hours by spraying frequently or by applying a surface covering or clean wet cotton matting, building paper, or other material. Since concrete in early stages is easily stained, make sure coverings are clean and free from rust, grease, tar, other foreign materials, and soluble colors that might leach onto surface. When artificial heat is used in the building, assure that surfaces of a freshly poured slab subject to direct heat from stoves or air ducts are kept wet.

*k. Waxing.* After floor has been in use about 30 days, apply one coat of colored wax to intensify the color and to offset any efflorescence which has not disappeared.

## 31. Repair of Concrete Floors for Heavy-Duty-Traffic

When concrete floors become rutted, spalled, or broken, replace the entire floor or resurface it partially. Use heavy-duty cement-mortar concrete inlay to resurface floors subjected to considerable oil dripping. If there is no oil drippage and the area cannot be kept clear of traffic for the 3 to 7 days needed for curing, resurface with asphalt mastic. If a concrete floor surface remains smooth and the only difficulty is dusting or excessive wear, use a liquid floor hardener (par. 36). Minor repairs to concrete treads are best accomplished by use of epoxy concrete (pars. 30 and 33) and other proprietary products in accordance with manufacturers' instructions.

## 32. Mud-Jacking

Concrete floor slabs on the ground, that have settled due to careless compaction of the subgrade, may be leveled by mud-jacking or similar treatment. A detailed discussion of filling voids under concrete slabs and raising settled slabs is given in TM 5-624.

## 33. Cement-Mortar Concrete Inlay

*a. Preparation of Floor.* Prepare a floor for cement-mortar inlay as follows:
(1) *Cutting.* Cut out defective areas of existing floor to a depth of at least 2 inches (fig. 24). Cut the whole area without leaving small islands of the original surface.

*Figure 24.   Preparing faulty concrete floor for cement-mortar inlay.   Note that entire
defective area has been cut out, with no islands left.*

If removing the slab down to fill is desirable, carefully tamp any disturbed fill. Keep edges around cut-out area as straight and vertical as possible. To avoid even minor featheredging along joint lines, give edges a final dressing by hand-chiseling if necessary. If a mason's power saw is used to block out areas to be repaired no edge trimming will be required. When a saw is used, first block out the area by sawing to a depth of 1½ inches to 2 inches, then remove the concrete on inside of lines to the desired depth.

(2) *Cleaning.* Remove all dust, dirt, loose particles, and other debris from cut by vigorous brooming, air-blasting, or other means.

(3) *Wetting.* About 4 hours before placing concrete inlay, saturate prepared surface thoroughly with water. Remove all surface water with a mop so surface can dry for about 1 hour before placing concrete inlay. If old slab is removed down to fill, place a layer of saturated felt or waterproof building paper on fill and omit wetting.

(4) *Cement-paste slush coat.* About 15 minutes before concrete inlay is placed, apply a slush coat of cement and water mixed to consistency of thick cream. Brush it well into surface with a stiff broom. Remove all excess cement paste leaving barely enough to give surface a uniform cement color. Omit slush coat if old slab is removed down to fill.

(5) *Thickness of concrete inlay.* If base for concrete inlay is chipped out so new inlay finishes level with existing floor, chipped-out section must be at least 2 inches deep. If old slab is removed down to fill, lay new slab to full depth of old one.

b. *Materials.*
(1) *Cement.* When floors must be put in service without delay, use high-early-strength cement meeting requirements for type III of Federal Specification SS-C-192, cement; portland, high-early-strength. Otherwise, use cement meeting requirements, type I or type II, of Federal Specification SS-C-192, cement; portland, with 2 percent of calcium chloride by weight.

(2) *Aggregates.* For aggregates use quartz, **trap** rock, granite, or other mineral of equivalent hardness, able to meet the Los Angeles abrasion test and other requirements, except grading, listed in Federal Specifications SS-A-281, aggregate for portland cement concrete. Grade coarse and fine aggregates within the following limits.

|  | Sieve size | Percentage passing sieve |
|---|---|---|
| Sand (Fine Aggregates) | ¼ inch | 100 |
|  | No. 4 | 90 to 100 |
|  | No. 8 | 70 to 90 |
|  | No. 16 | 60 to 80 |
|  | No. 30 | 40 to 60 |
|  | No. 50 | 5 to 15 |
|  | No. 100 | 0 to 3 |
| Rock (Coarse Aggregates) | ⅜ inch | 100 |
|  | No. 4 | 65 to 85 |
|  | No. 8 | 40 to 60 |
|  | No. 16 | 0 to 15 |
|  | No. 30 | 0 to 2 |

If above requirements cannot be met with available local materials, use following grades:

|  | Sieve size | Percentage passing sieve |
|---|---|---|
| Sand (Fine Aggregates) | ¼ inch | Not less than 100 |
|  | No. 50 | Not more than 15 |
|  | No. 100 | Not more than 3 |
| Rock (Coarse Aggregates) | ½ inch | Not less than 100 |
|  | ⅜ inch | Not less than 95 |
|  | No. 16 | Not more than 5 |

(3) *Water.* Use potable water for mixing.
(4) *Bonding agents.* Various proprietary products are marketed as agents to promote bonding of new concrete to old concrete. The effectiveness of such bonding depends on the proper preparation of the surface and the curing of the new concrete. When inadequate or borderline practices are followed, unsatisfactory results are likely to be obtained whether a bonding agent is used or not. Liquid polysulfide epoxy adhesives possess unusual bonding qualities for bonding new concrete to old. They should be used according to manufacturers printed instructions.

c. *Proportions.* Use 1 part cement, 1⅓ parts sand, and 2 parts rock by volume. Use not more than 4½ gallons of mixing water for each sack of cement including free water in the aggregate, for hand finishing, and not more than 4 gallons for machine

*Figure 25. Completed concrete inlay. White line indicates calking in expansion joint.*

finishing. These proportions make a fairly harsh mixture which requires more than ordinary effort to finish. Proportions given are necessary, however, to prevent shrinkage and get a good wearing surface. If local materials mixed in these proportions yield a concrete too harsh for satisfactory finishing, change the proportions to 1 : 1½ : 1½.

*d. Mixing and Placing.* Mix each batch mechanically 2 to 3 minutes, place material on slab, and vibrate, roll, or tamp it firmly in place. A grill type tamp can be made by nailing strips about ½-inch wide and ½-inch deep, spaced about ⅜-inch apart, to the face of an ordinary tamp. Screed material to designated levels and float surfaces with wood or, preferably, a power float. Let stand for 30 to 45 minutes or until pressure from finger ceases to make a dent, then steel-trowel to final finish. Do not dust dry cement or cement-sand mixture, or sprinkle water on it in finishing. Be careful that finishing does not bring any excess fines to the top. Maintain existing expansion joints.

*e. Curing.* Place a suitable covering material or curing compound on the new surface as soon as it has set enough so surface will not be marred. A completed inlay is shown in figure 25.

*f. Re-Use of Floor.* If high-early-strength cement is used, the floor can be used after 3 days unless temperature has been below 45° F. In that case allow 5 days. Where standard portland cement is used, do not use the floor for traffic loads for 7 days after installation.

## 34. Heavy-Duty Concrete Topping

Repair heavy-duty concrete floors that begin to dust and ravel by installing a 2-inch thick heavy duty concrete topping over existing concrete (fig. 26). Control joints in new topping should coincide with those in existing floor. Aggregate should comply with Federal Specifications SS-A-281, except the gradation of coarse aggregate shall be:

| Coarse aggregate | |
| --- | --- |
| U.S. standard sieve size | Percent by weight passing each sieve |
| ¾ in. | 100 |
| ½ in. | 90 to 100 |
| ⅜ in. | 40 to 70 |
| No. 4 | 0 to 15 |
| No. 8 | 0 to 5 |

Roughen and clean existing floor. This method may

also be used in constructing new heavy-duty concrete floors. When used, construct base course of required thickness as for floor slabs (pars. 29 and 30), and rough finish (fig. 26). Place 2-inch topping when concrete is sufficiently hard, as outlined above.

## 35. Heavy-Duty Mastic Topping

*a. General.* To cover existing concrete floors, use a combination of cement, asphalt emulsion, water, and sand, with optional additions of suitable stone mixes. When hard, cured, and ready for use this mastic topping forms a tough, resilient, dustless composition, highly resistant to abrasion and shock from warehousing vehicles.

*b. Thickness.* Finish mastic topping to a thickness of at least ½ inch above the highest section of existing floor unless additional thickness is required. Feather all free edges.

*c. Preparation of Concrete Floor Surfaces.*

(1) *Cleaning.*

(a) Clean all dust, dirt, and foreign particles from concrete-floor surfaces by vigorous brooming, air-blasting or other means.

(b) Remove oil and grease spots with a solution of 1 gallon of hot water and ¼ pound of common household lye or trisodium phosphate. Apply hot solution to the spot and allow it to soak about 10 minutes, then scrub with a stiff bristle, fiber, or wire brush. Mop and flush with clean water, continuing this process until cleaning compound is removed.

*Note.* Take care to keep cleaning compound away from skin and eyes.

(2) *Priming.* While cleaned concrete floor is still damp but not wet, apply asphalt-emulsion primer in one or two coats as directed by manufacturer (fig. 27). If two coats are used, dilute first coat with about 20 percent clean, cool water. Scrub it over entire concrete surface, using at least 1½ gallons for 100 square feet of surface, and allow time for it to become tacky dry. Apply second coat of primer full strength, at least 1 gallon for 100 square feet of surface, and allow it to become tacky dry before placing the finish mastic topping. To strengthen the mechanical bond, make the second prime coat one part primer to one part sand and scrub it thoroughly over all surfaces.

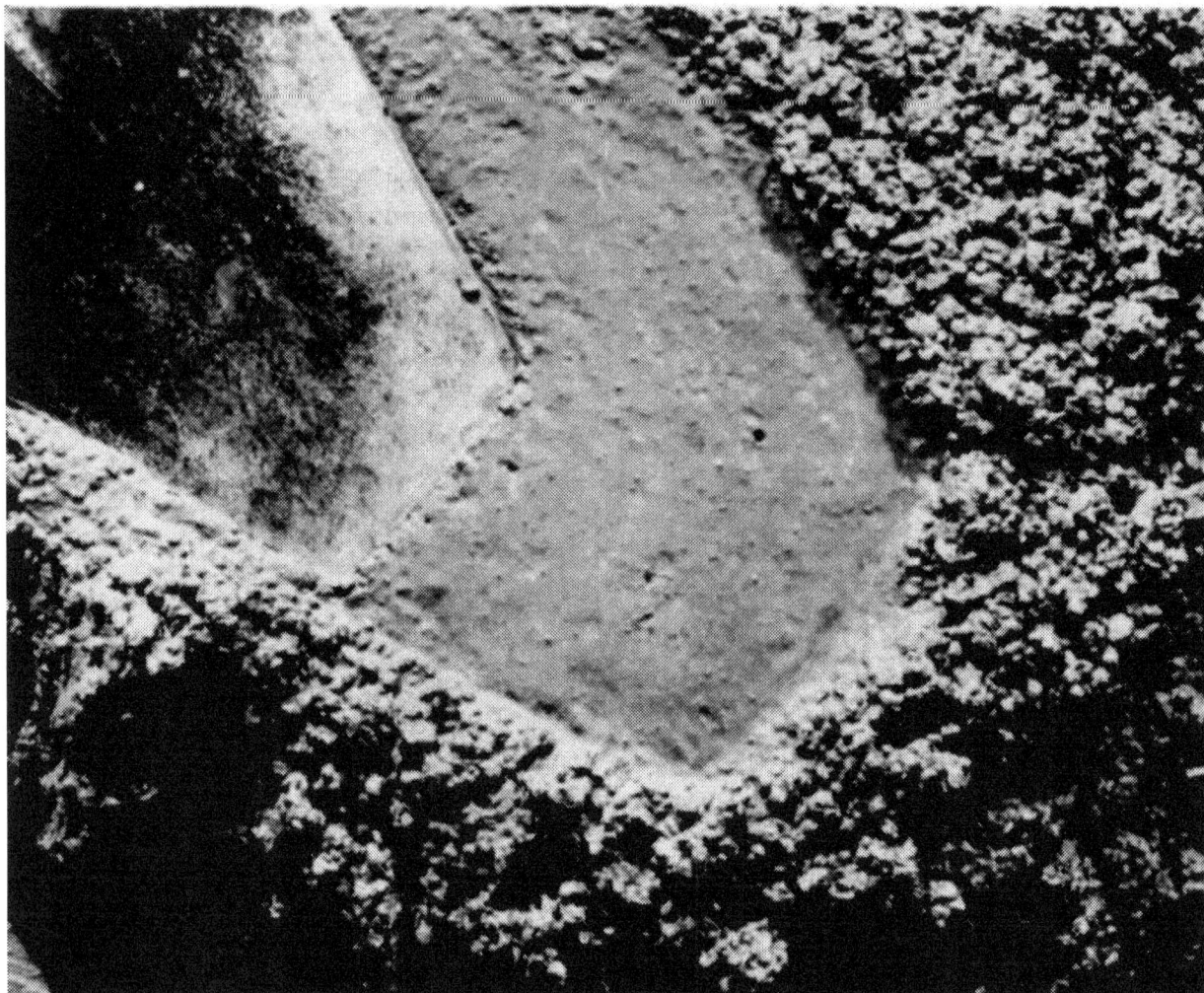

1  A properly proportioned mix

*Figure 26.  Concrete floor finish construction.*

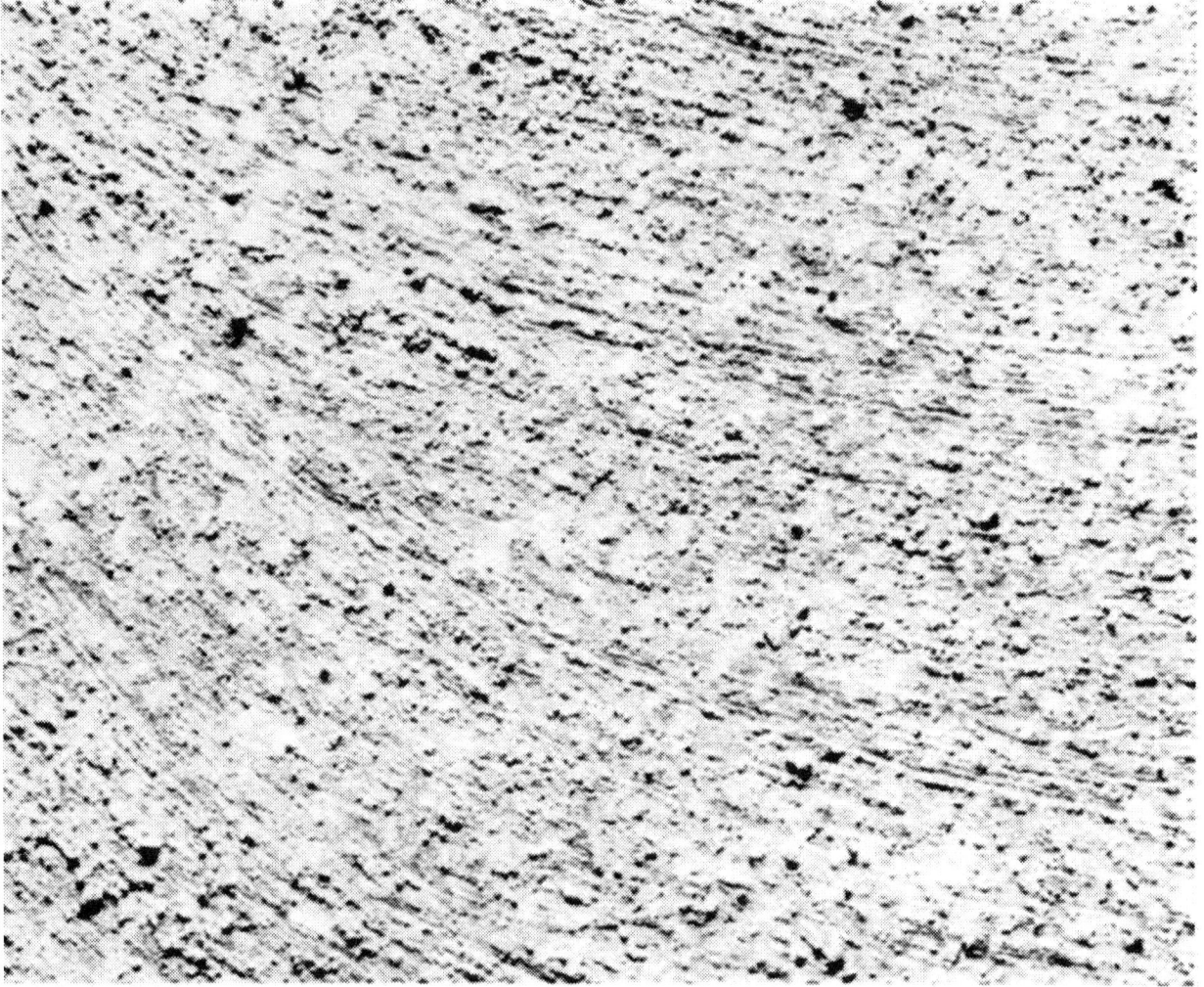

2  Base finish for wearing course

*Figure 26*—Continued.

3   Development of wearing course

*Figure 26*—Continued.

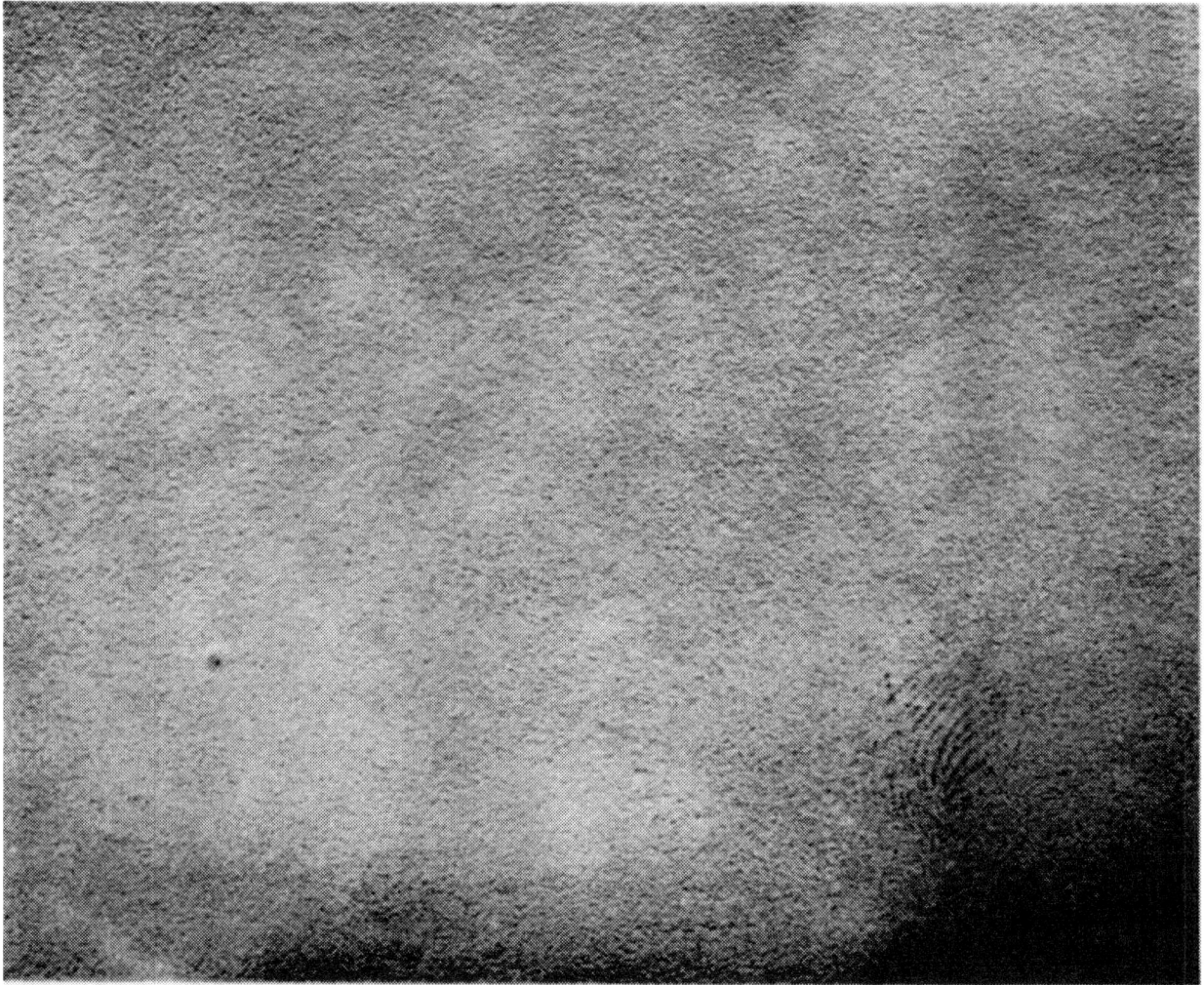

4  Properly troweled finish

*Figure 26*—Continued.

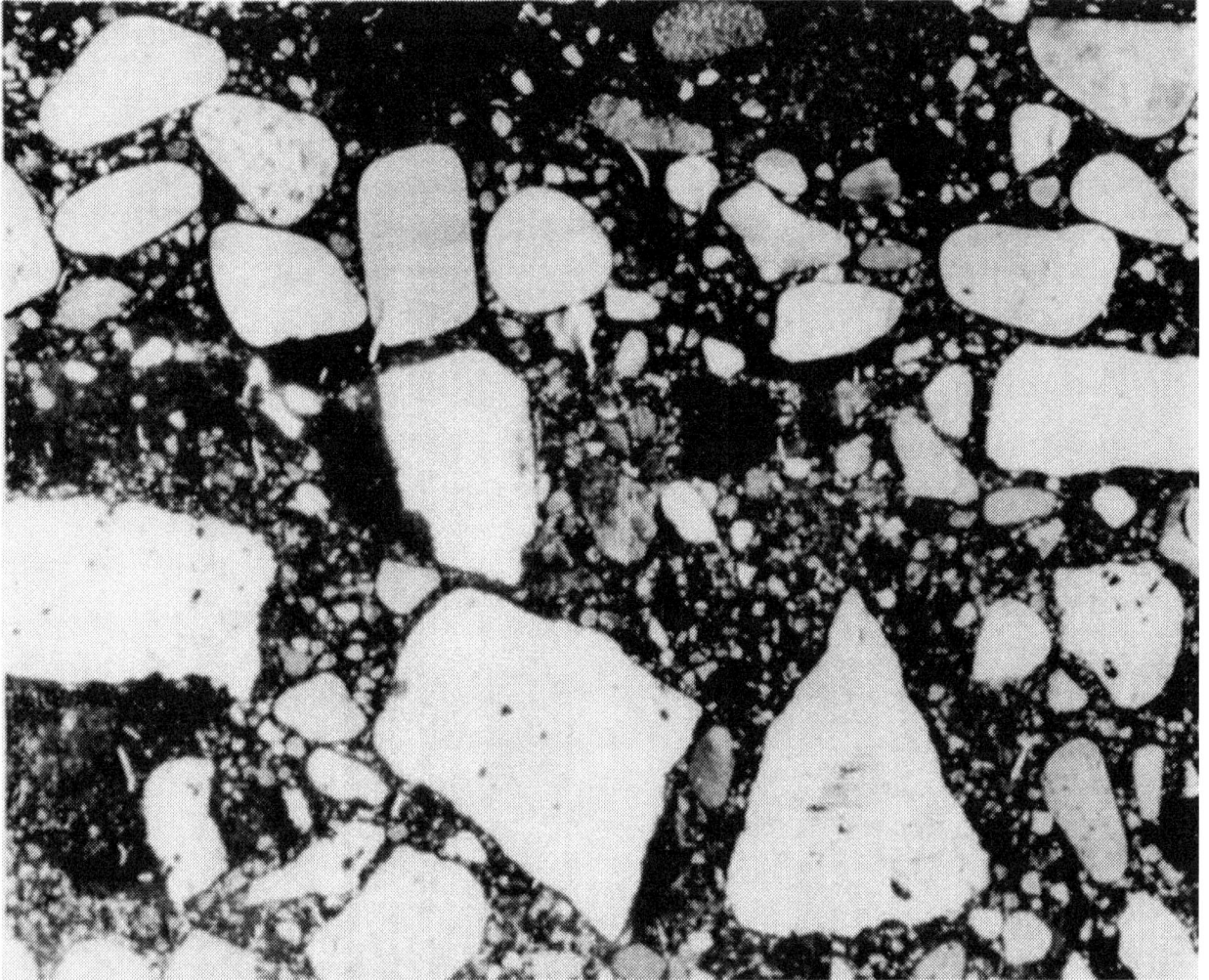

5   Cross-section of properly constructed concrete floor

*Figure 26*—Continued.

6   Cement and sand mortar finish

*Figure 26*—Continued.

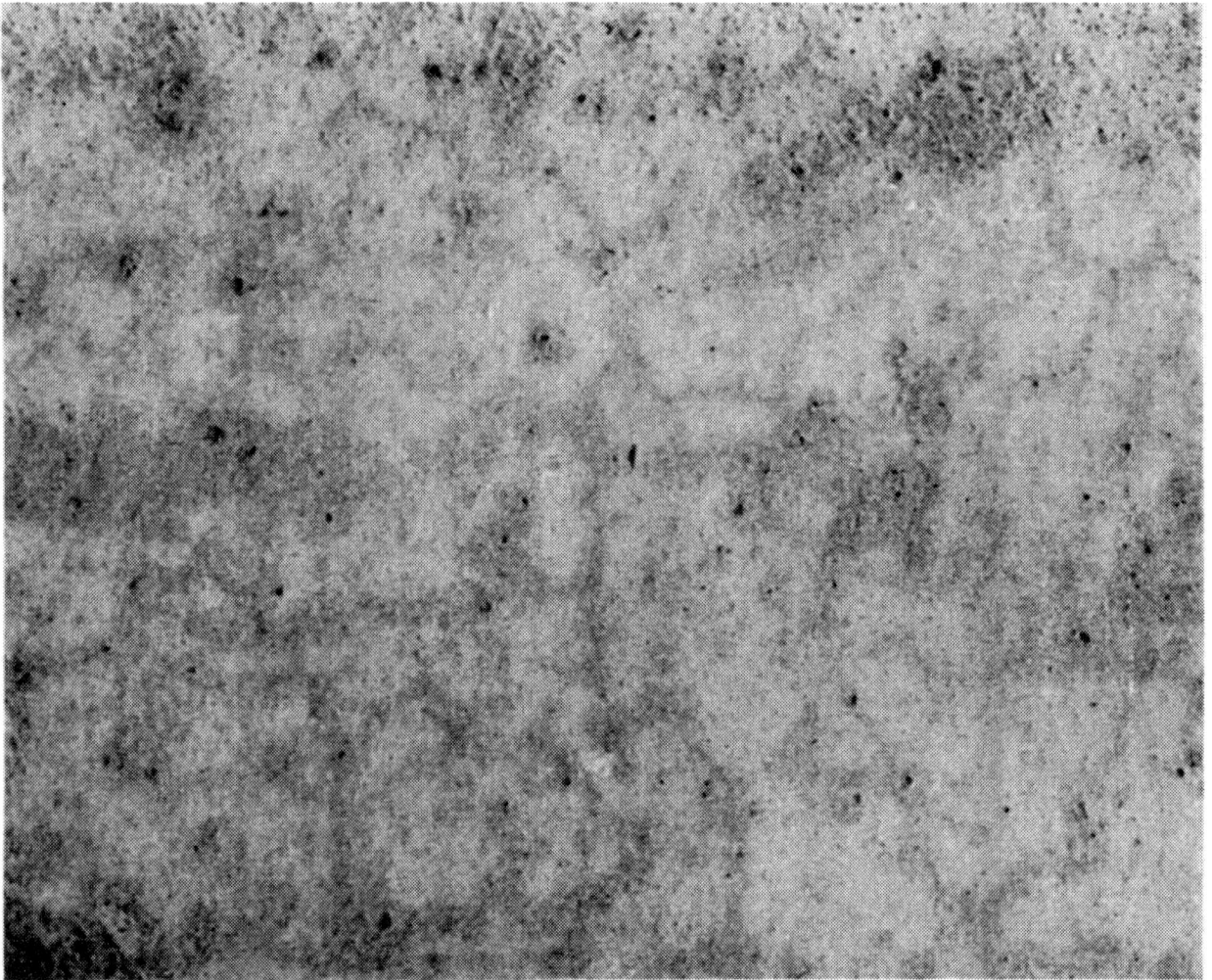

7   Surface crazing

*Figure 26*—Continued.

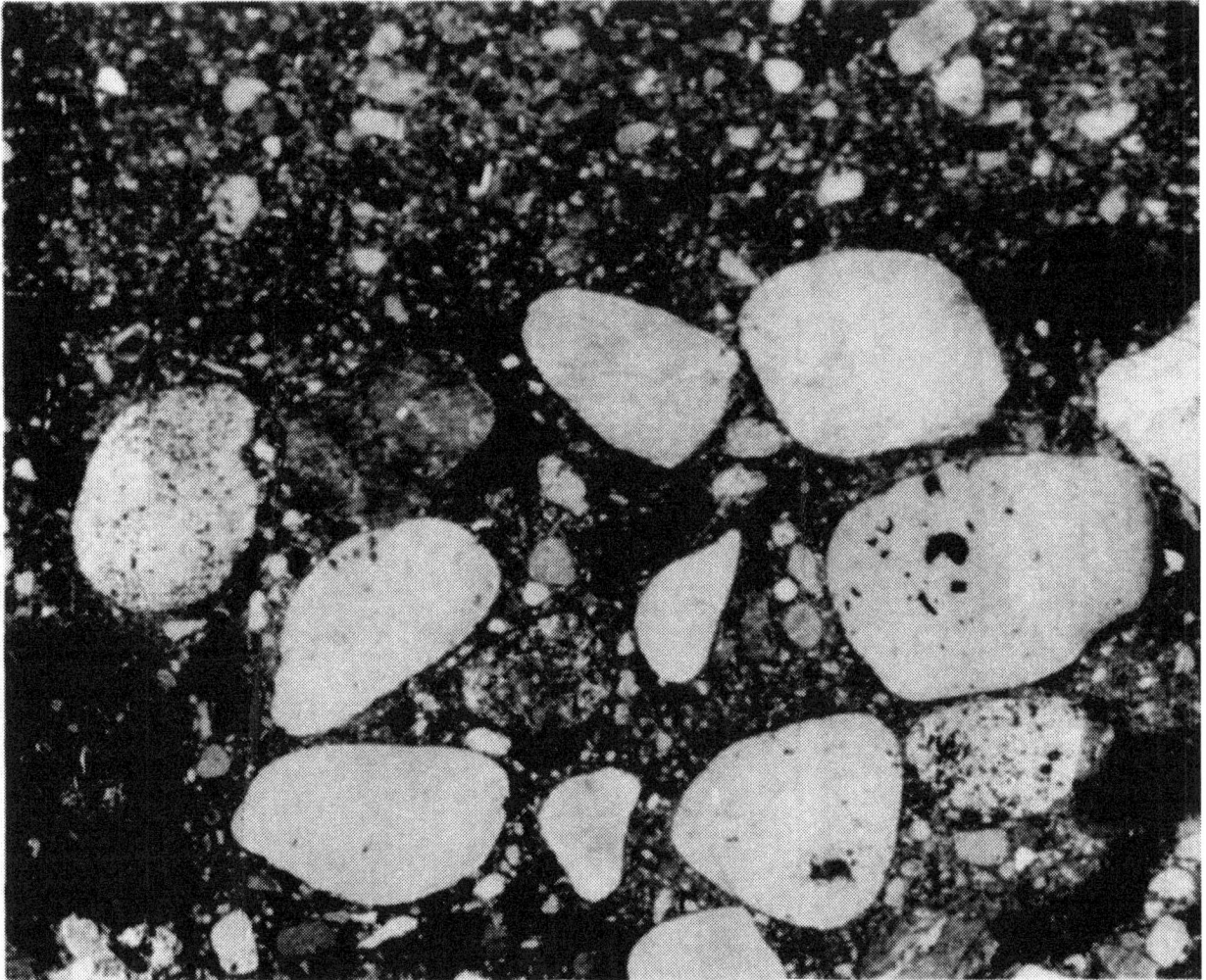

8 Cross-section of improperly constructed concrete floor

*Figure 26*—Continued.

*Figure 27. Applying primer to cleaned and dampened concrete.*

*d. Topping Materials.* Use topping materials which conform generally to the following requirements:

(1) *Cement.* Federal Specification SS–C–192, cement, portland, type I or type II.

(2) *Asphalt emulsion.* Emulsion should be manufactured by a company having a long period of success with this material in compounding bitumen toppings on heavy duty warehouse floors, and outside concrete ramps, and should be a standard advertised and stocked item of that firm. It should be either collodial clay or chemical type, composed of a uniform dispersion of asphalt in water and an emulsifying agent which gives uniform results.

(3) *Sand.* Sand should be clean, sharp, free from loam, organic matter and other impurities.

(4) *Aggregate.* Aggregate should be gravel, crushed stone, or granite chips, ⅛- to ⅜-inch size. Do not use aggregate where featheredging is required.

(5) *Water.* Water should be potable.

*e. Proportions.*

(1) *For hand mixing* use the following mixtures:

(a) *Mixture 1.* One part portland cement, two parts asphalt* emulsion (clay type), two parts sand, and three parts coarse aggregate by volume. To put a ½-inch topping over 120 to 130 square feet of surface, use 1 bag portland cement (1 cubic foot), 15 gallons asphalt emulsion (clay type), 2 cubic feet sand, and 3 cubic feet coarse aggregate.

(b) *Mixture 2.* One part portland cement, two parts asphalt* emulsion (clay type), and 3½ parts sand by volume. The mixture can be used where coarse aggregate is not available or where featheredging is required. To cover 100 square feet with ½-inch topping, use 1 bag portland cement, 15 gallons emulsion (clay type) and 3½ cubic feet sand.

(2) *Machine mixing.* Use the following mixture (mixture 3) for machine mixing and finishing: One part portland cement, two parts asphalt emulsion (clay type), two parts sand (torpedo type), and six parts coarse aggregate by volume. Using this ratio, 1 bag portland cement, 15 gallons emulsion (clay type), 2 cubic feet sand, and 6 cubic

feet coarse aggregate will provide ½-inch topping over 170 to 175 square feet of surface. Reduce quantity of emulsion to 1½ parts for chemical type. Better performance will be achieved for exterior use, if mixture is proportioned as follows:

(a) If chemical-base type emulsion is used:
1 part portland cement
1½ parts asphalt emulsion
2 parts sand
3½ parts coarse aggregate.

(b) For industrial clay-type emulsion use mixture No. 1 above. Smooth gravel is not recommended for coarse aggregate for exterior mastic toppings as it tends to become exposed under heavy concentrated wheel loads. Prime clean concrete with a mixture of one part water and one part emulsion.

*f. Mixing.*

(1) To prepare mixtures 1 and 2, mix measured quantities of cement, sand, and coarse aggregate by hand. Add enough water to make a heavy mortar. Add emulsion and keep mixing until whole mass is uniformly black.

(2) For machine mixtures use a batch rotating-drum concrete mixer or a mortar mixer (stationary, tilting, open-top drum with revolving blades). Run mixer 4 to 5 minutes to mix a cubic-yard batch thoroughly.

*g. Laying.* Place ½-inch or thicker screed strips on the concrete floor to control areas and depth of topping (fig. 28). After mix has been rough-placed, level with a straightedge strike-off board (fig. 29). After initial set has taken place, woodfloat topping to make the surface level. Initial set occurs 4 or 5 hours after installation; final set, in about 12 hours.

*h. Tamping and Rolling.* Densify the installed mix with a tamper, a roller, or both. Use a tamper made of perforated steel plate or heavy wire mesh, two meshes to the inch. Use a roller about 30 inches wide, weighing at least 10 pounds for each inch of width. Do not tamp or roll the topping until it is too hard to be picked up by the tamper or to show footprints following the roller.

*i. Power Floating.* A standard power-floating machine makes the topping denser. Always use a power-floating machine on mixture 3. It can also be used on mixture 1 and 2, except on featheredging where hand-troweling is required.

---

* *Note.* If chemical emulsion is used only 1½ parts are required.

*Figure 28. Placing asphalt mastic mix between screeds.*

*Figure 29. Striking off asphalt mastic with straightedge.*

Figure 30. Steel-troweling asphalt mastic. Edges broken down with trowel for featheredging.

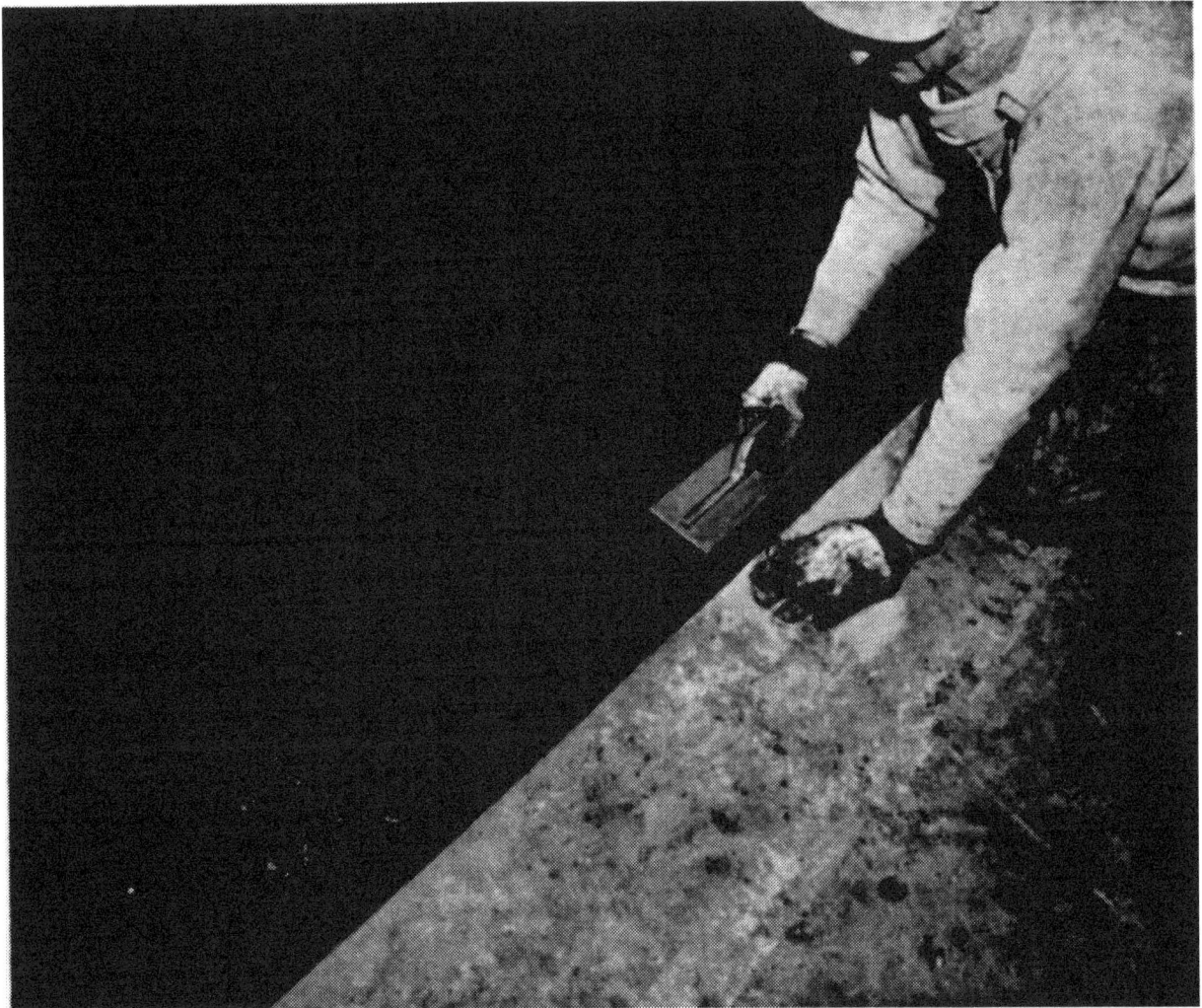

*Figure 31. Forming the feather edge.*

Figure 32. Completed asphalt mastic trucking strip.

*j. Steel-Troweling.* Steel-troweling can be used after tamping, rolling or power floating to give a smoother finish (figs. 30, 31, 32).

*k. Curing.* Keep finished topping damp for about 12 hours by any suitable method. Too rapid drying may cause slight checking, which will disappear under action of wheeled traffic.

*l. Advantages.* Heavy-duty mastic topping has the following advantages:

(1) Easily and quickly laid.
(2) Ready for use in the shortest possible time.
(3) Dense enough to permit trucking from the time topping is put into service.
(4) Economical to maintain.
(5) Waterproof, noise-deadening, and dustless.

*m. Costs.* Depending on materials used, area covered, and local conditions affecting the work, the total cost of a heavy-duty ½-inch thick asphalt mastic topping, including materials and labor, varies from 25 to 40 cents a square foot.

## 36. Liquid Hardeners

Liquid hardeners (par. 30*g*) applied over friable, dusting concrete floors harden the concrete and protect the surface against abrasion and shock caused by heavy traffic. Use low-viscosity solutions which penetrate deeply into the concrete.

*a. Preparation of Concrete Floor Surfaces.* Clean all dust, dirt, foreign particles, and oil or grease spots from concrete floors to which a liquid hardener is to be applied.

*b. Materials.* Use one of the following materials or an approved equivalent:

(1) *Sodium Silicate (Water Glass):* Commercial, 40° to 42° Baume (Be).
(2) *Magnesium Fluosilicates:* Crystalline salts plus zinc fluosilicate.

*c. Mixing and Applying Sodium Silicate.* Dilute sodium silicate just before using it by adding 4 gallons of water to 1 gallon of sodium silicate. One gallon of this solution covers about 800 square feet of floor surface with one coat. Apply it as follows:

(1) *First coat.* Apply first coat with a mop or broom, brushing solution continuously over floor surface to get even penetration. Allow at least 24 hours for first coat to dry and harden. Scrub dried surface with hot water to remove the glaze which generally appears, then allow 24 to 48 hours for surface to dry completely. If floor is porous enough to absorb first coat without leaving a glaze, hot-water scrubbing can be omitted.

(2) *Second coat.* Apply a second coat in the same manner as the first. Allow it to dry, scrub with hot water, and again dry for 24 to 48 hours.

(3) *Third coat.* Apply third coat in the same manner. Allow it to dry thoroughly before using floor. Hot-water scrubbing is not needed on this coat.

*d. Mixing and Applying Magnesium Fluosilicate.* To 1 gallon of fresh, clean water add a 2-pound mixture of crystalline salts of magnesium fluosilicate and zinc fluosilicate, at least ½ pound being zinc fluosilicate. Fluosilicates can be obtained as prepared solutions or as dry crystals. Dry crystals are more economical and can be mixed at the job with fresh water in wooden vessels. One gallon of this solution covers about 100 square feet of floor surface with one coat. Apply it as follows:

(1) *First coat.* Dilute the solution by adding 1 gallon of fresh, clean water to each gallon of solution, or prepare a new solution of 1 pound of dry crystals to 1 gallon of water. Apply solution with a mop or broom and brush it continuously over floor surface for several minutes to get even penetration. Allow at least 24 hours for drying before applying second coat.

(2) *Second coat.* Apply second coat, using undiluted solution of 2 pounds of crystals to 1 gallon of water. Follow same procedures as for the first coat. Allow it to dry for at least 24 hours.

(3) *Third coat.* If floor is unusually friable or porous, apply a third coat, using same solution and application as for second coat.

*e. Costs.* Average cost of treatment using three coats either of sodium silicate or magnesium fluosilicate is about $0.02 a square foot. Sodium silicate materials are cheaper than fluosilicates, but labor cost is greater because of the hot-water scrubbing of each undercoat.

*f. Comparative Effectiveness.* Laboratory experiments and actual use have generally shown little difference in effectiveness between water glass and fluosilicate if both are applied properly. On smooth-troweled floors or dense surfaces, fluosilicate treatment may be better because fluosilicate solution is less viscous and penetrates more easily.

(1) Penetration of most hardeners, especially sodium silicate, is improved by applying the solution heated to about 170° F. Both sodium silicate and fluosilicate solutions

become about half as viscous at 170° F. as at 70° F. This change in viscosity is greater than that obtained by diluting the solution, which is sometimes recommended for the first coat.

(2) Maximum penetration of hardener may not always be desirable. With rough porous surfaces it is sometimes desirable to retain enough hardener at the surface for adequate reinforcement, and use of more concentrated hardeners applied cold will probably give better results. On rough or porous surfaces, 42° Baume sodium silicate solution diluted 1 to 3 or 1 to 2.5 is preferable to fluosilicate solution.

(3) Most liquid hardeners sold under trade names have a base of sodium silicate or magnesium fluosilicate, with a small amount of zinc fluosilicate, zinc sulfate, or other materials sometimes added.

(4) Sodium silicate and magnesium fluosilicate, with or without added materials, are favored because of their availability, ease of mixing and application, ability to harden surfaces satisfactorily, and comparatively low cost.

g. Sealer. The material covered by Federal Specification TT-S-176, Class 2, is a satisfactory sealer for concrete floors subjected to heavy traffic that have not been treated with a hardener or painted.

# SECTION VII

# MAINTENANCE AND WATERPROOFING OF CONCRETE WALLS

(See section XI also)

## 37. General

The guidance provided in paragraphs 19d, 40-42, 61 and 62 is applicable also to maintenance and waterproofing concrete walls. Normally, concrete walls require a minimum amount of maintenance. Cut out cracks to a depth of one inch to a width of at least ½ inch. Wet down before filling with a 2 to 1 sand-cement mortar. Repair large broken areas by cutting out sufficiently to expose reinforcing rods or mesh that will bind the new concrete to the remaining wall (fig. 33). Thoroughly clean and wet the cut surfaces of the wall then coat with a slurry of neat cement and water as described in paragraph 33a, prior to placing the new material. Mix the new concrete according to the requirements of paragraphs 18–22. Most defects that cause appreciable problems, such as leakage are due to the expansion and contraction of the building members caused by temperature changes. Lack of adequate expansion and contraction joints is a common cause of cracking (fig. 34). Other common causes are settlement, poor materials, structural weaknesses in the foundation, excessive floor loadings, and poor workmanship in original construction. Efflorescence on walls always indicates trouble. It usually appears as a light powder or crystallization on evaporation of water. Excessive moisture in the walls may be caused by defective flashings, gutters, downspouts, copings, or improperly filled joints. Location of efflorescence does not always mean that water is entering the wall at that point. Streaks on the wall from the top down or patches some distance from the top might point to defective gutters or copings. Patches of efflorescence are sometimes caused by opened joints. Water may also enter openings at window sills or around window and door frames. Efflorescence close to the ground may indicate ground water drawn up by capillary action. Open texture concrete walls may be made more durable, more pleasing in appearance, and exclude entrance of moisture by painting with portland cement paint as described in paragraph 61b(2).

## 38. Cracks

Horizontal movement cracks are usually long wide cracks that occur along the line of the floor or roof slab, or along the line of lintels over windows. Vertical and diagonal movement cracks generally occur near the ends or offsets of buildings. They may also be found extending from a window sill to the lintel of a door or window on a lower floor. These vary from ⅛ to ⅜ inches in width. Do not use brittle materials to repair such cracks; use a flexible compound (TM 5–620). A good nonhardening, nonshrinking caulking material is Polysulfide sealing compound complying with Federal Specification TT–S–00227 (Class A, self-leveling; Class B-non-sag.) The cost is about $20.00 per gallon, but since

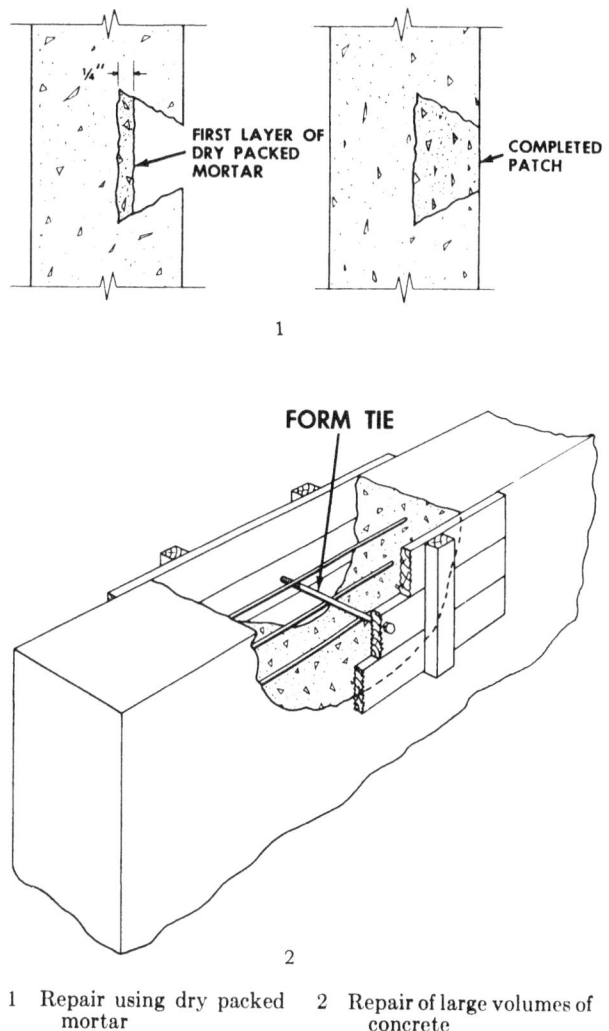

1

FORM TIE

2

1  Repair using dry packed mortar    2  Repair of large volumes of concrete

*Figure 33.  Wall repair.*

FORM SHEATHING    "V" JOINT

BULKHEAD

3

SECTION OF WALL ALREADY POURED

KEYWAY

BEVELED 2" x 4"

BULKHEAD 1" BOARDS

CLEATS

1" BOARDS

SPREADER BLOCKS

NUTS & WASHERS

1" x 6" S

2

$\frac{1}{3}$ TO $\frac{1}{4}$ T    T

4

REINFORCING BARS

JOINT FILLER    METAL WATER STOP

5

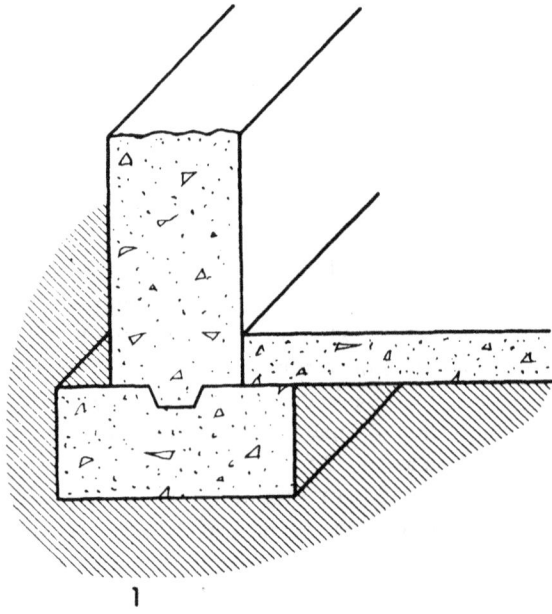

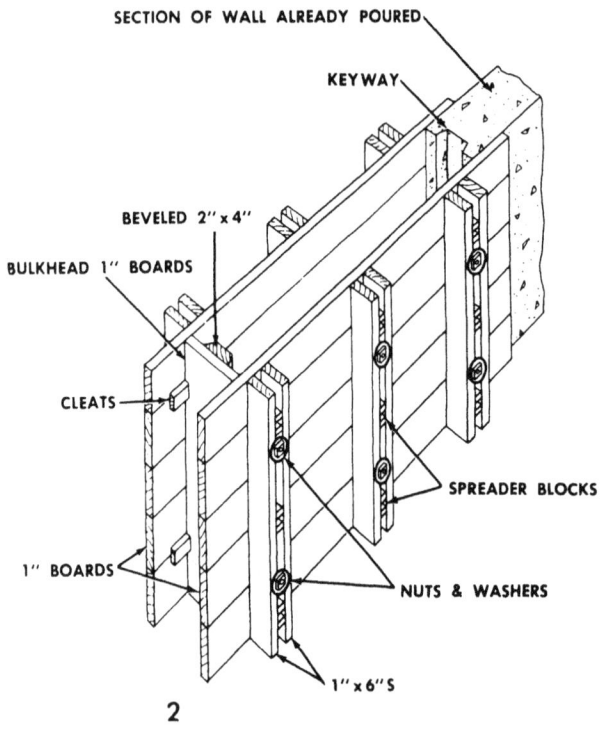

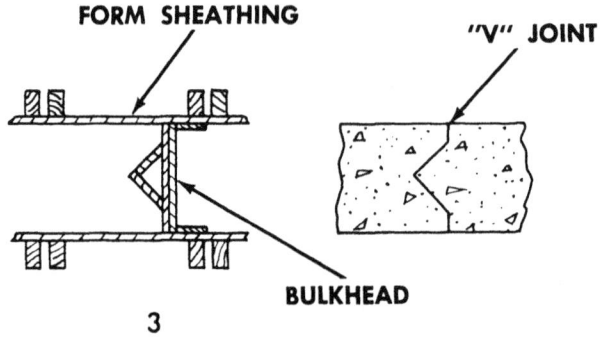

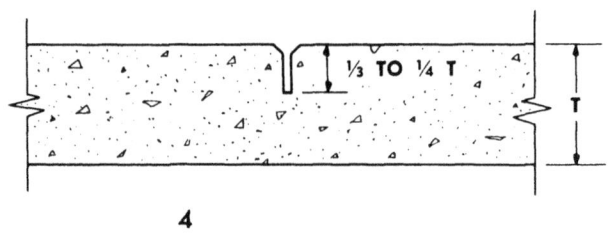

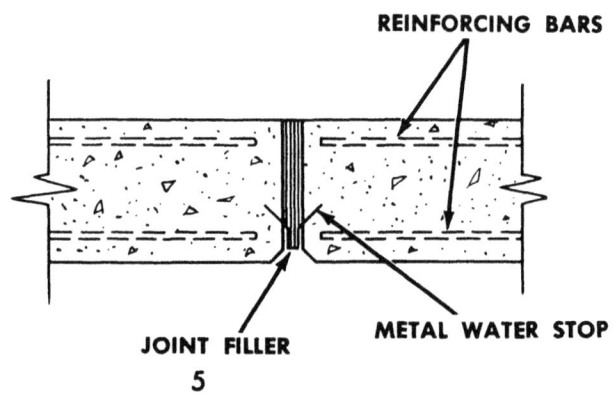

| 1 | Joint between wall and footing | 2 | Vertical bulkhead in wall | 3 | Vertical V-construction joint | 4 | Dummy contraction joint |
|---|---|---|---|---|---|---|---|
| | | | | | | 5 | Expansion joint for wall |

*Figure 34. Contraction and expansion joints.*

it has a long useful life, it is relatively inexpensive on an annual cost basis. Shrinkage cracks are fine hairlike cracks. In repairing shrinkage cracks, do not chisel out and fill; scrub with a grout made with 65 percent cement and 35 percent sand (100 percent passing the No. 50 sieve) mixed to consistency of heavy molasses. Wet the wall and scrub in. Cure with water. Postpone shrinkage crack repairs until wall is at least 1 year old. In special cases, as when the concrete is exposed to severe weathering or

AGO 10179A

corrosion, (structure exposed to sea water), repair broken or spalled concrete as soon as practicable to prevent progressive deterioration which might result from rusting of reinforcing steel. Seal cracks to prevent ingress of water that would promote such corrosion and would also subject the concrete to danger of further deterioration by freezing and thawing. Pneumatic application of mortar may often be employed advantageously in such repair work (par. 25c(4)(c)).

## 39. Spalled and Eroded Surfaces

a. *Preparation of Surfaces.* Treat spalled and eroded surfaces that cannot be renewed by brush coats of thick cement water paint (par. 61b(2)), by:

    (1) Removing all loose and fractured surface material by use of hand chisel or air hammer.

    (2) Cleaning and repairing all exposed steel.

    (3) Roughening remaining smooth surfaces by use of wire brush or sand blasting.

b. *Plastering.* Prior to beginning plaster work patch all deep recesses as outlined in paragraph 38. After all patches have attained initial set, saturate the surface with water for 1 to 2 hours preceding plastering. Plaster should be composed of 1 part of portland cement; 2 to 2½ parts of masonry sand; and 10 percent hydrated lime. Add a water repellent material if desired (pars. 62m and n(3) and 61a(3)). Apply plaster in ⅜″ to ¾″ thick layers in the usual manner, and cure. If surface is only partially plastered and uniformity in color and waterproofing is desired, paint with cement water paint (par. 62f and n).

c. *Pneumatically Applied.* In lieu of plastering, surfaces may be renewed by pneumatically-applied material (par. 25c(4)(c)). This type of work is usually done by specialists. In some instances, best results are obtained by anchoring reinforcing mesh to receive the material. It is an excellent method of repairing concrete water tanks.

# REPAIR AND WATERPROOFING OF UNDERGROUND STRUCTURES

## 40. General

*a.* Many of the materials, methods, and principles required for the repair of underground structures are similar to those for maintenance and waterproofing masonry walls (pars. 60–62), and maintenance and waterproofing concrete walls (pars. 37–39). Appendix IV gives the bituminous-membrane method utilized for general waterproofing problems. These principles may be applied to most underground concrete structures. The underground ammunition igloo has resulted in a particular waterproofing problem; therefore, this section deals only with such problem and describes a special treatment for ammunition igloos.

*b.* Examine carefully underground ammunition igloos reported to require repair due to leakage, to establish beyond a doubt that the undesirable conditions are caused by leakage rather than condensation or other causes. Modifications to storage arrangements, improved ventilation, or both may correct problems due to condensation.

## 41. Preparation for Waterproofing

Remove the earth cover from the entire structure to the ledge of the foundation footings including the rear end wall and stockpile for replacement, (fig. 35). When removing the earth, be careful not to damage the concrete structure or the drain tile near the

*Figure 35. Uncovering typical igloo.*

*Figure 36. Patching concrete surface.*

footings. Scrape the concrete structure to remove all dirt, foreign matter, and any loose adherent material. Remove as much as practicable of the old waterproofing layer if the new waterproofing is not compatible. If compatible only the portions of the old waterproofing that are damaged or loosely adherent need be removed. Inspect the cleaned surfaces; remove all sharp edges, ridges, abrupt cleavages, bolts, and projections; fill low spots, deeply honeycombed areas, marked depressions, holes, and severe roughness with portland-cement grout, thoroughly bonded to the old concrete, trowelled smooth, and cured. Remove concrete pedestals and grout the resulting depressions (fig. 36). Clean all cracks, fill with asphalt-asbestos plastic cement and cover with two layers of 30-lb. roofing strips not less than 6 in. wide, embedded in two layers of asphalt plastic cement ⅛ inch in thickness, centered over each crack (fig. 37). Cracks ¼ inch or more in width shall, in addition to above, be covered with sheetmetal strips 7 inches wide, sandwiched between the two roofing strips 8 inches wide and heavily embedded in asphalt plastic cement. Cracks ½ inch and larger should be filled with a mix composed of 8 parts asphalt emulsion and two parts of portland cement (8 : 2 mix) prior to installing felt strips. Seal the horizontal joint between the foundation and barrel also with this 8 : 2 mix (trowelled into the joint and spread 18 inches in width). Firmly embed a 12-inch wide strip of glass fabric centered over the joint, and coat with a second application of the above mixture. Rough, pitted, honeycombed areas may also be filled with the 8 : 2 asphalt emulsion-portland cement mixture in lieu of cement grout where conditions warrant. When the concrete is found to be porous and saturated with water, dry out adequately to provide a completely

Figure 37. Applying felt over cracks.

dry surface before the prime coat is applied. Normally the drying out period will be about 14 days.

## 42. Waterproofing

a. *Materials.* Use fresh materials from original containers. Do not store or apply in freezing weather, or when frost is anticipated.

(1) *Primer.* The primer should comply with Federal Specification SS–A–701. Apply at the rate of not less than ½ gallon per 100 square feet.

(2) *Asphalt emulsion.* The asphalt emulsion should comply with ASTM Designation; D1187–51T, Type A. Apply at the rate of not less than 10 gal. per 100 square feet. Use additional material on rough areas. Do not permit adulteration or cut back of asphalt emulsion.

(3) *Glass fiber fabric.* The glass fiber fabric (Fed. Spec. HH–C–00466–GSA–FFS) should have a thread count of not less than 20 to the inch in direction of warp, not less than 10 in direction of fill, weigh not less than 1.85 ounces per square yard, have a tensile strength of not less than 75 pounds per inch of width in either direction, and should be coated with a compatible asphalt. Apply not less than two plies over the entire barrel and wall, foundation and footing ledge, and three plies over the horizontal joint between foundation and barrel.

(4) *Slip sheet.* The slip sheet material should conform to Federal Specifications SS–R–501 for Class A, 65 lb. per roll mica surfaced roofing. Apply slip sheet to covering the barrel, end wall, and foundation to the footing, with a center longitudinal lap of 24 inches and side laps of not less than 2 inches.

*Figure 38. Applying first coat over primed surfaces.*

(5) *Asphalt plastic cement.* The asphalt plastic cement should conform to Federal Specification SS-C-153, Type I.

b. *Application of Waterproofing Membrane.*

(1) *Area to be waterproofed.* Cover the entire structure with waterproofing including the barrel, the rear end wall, the front wall cant, the foundations to the footings, and at least 3 inches of the footing ledge.

(2) *Prime coat.* Inspect and repair as required to provide a reasonably smooth surface. Coat the entire surface to be waterproofed with primer applied at not less than ½ gallon per 100 square feet. Allow it to dry for at least 2 days.

(3) *First coating.* Apply asphalt emulsion over the prime coat at a rate of not less than 3½ gallons per 100 square feet (fig. 38) Apply two plies of glass fiber fabric, which will cover the entire area to be waterproofed over the fresh coating of asphalt emulsion (fig. 39). The fabric should be draped vertically over the barrel; and end laps, when required, should be not less than 12 inches and should be cemented over the lower ply. Side laps in each ply should not be less than 2 inches. Lay fabric vertically on the end wall, lapped at least 12 in. on the barrel for anchorage, without wrinkles and buckles; firmly pulled into the emulsion; smoothed by hand to work out wrinkles; and made tight to the lower walls for adhesion to prevent floating as the emulsion is applied.

(4) *Second coating.* Saturate the fabric completely by spraying asphalt emulsion at a rate of not less than 4 gallons per 100 square feet at approximately 80 psi pressure. Allow to dry for at least two days.

*Figure 39. Glass fiber membrane being applied.*

(5) *Third coating.* Apply the finish coat of asphalt emulsion over the entire area to be waterproofed, using not less than 2 gallons per 100 square feet, giving special attention to thin areas and areas in which the fabric is insufficiently coated. Inspect the coating after it has dried completely and touch up any bare spots or defects.

c. *Slip Sheet Protective Cover.* Dust the waterproofed area with talc or dry portland cement at the rate of not less than 1½ bags per igloo. Lay the slip sheet cover vertically with a lap of not less than 24 inches at the crown of the barrel and a lap of not less than 2 inches at the side (fig. 40). Sand bag the sheets to hold them in place at the crown lap until the earth cover is replaced.

d. *Repairs to Drain Tile.* Examine the drain tile and verify the following points:

(1) Tile is adequate in size.

(2) Highest point is one to two feet minimum below floor level.

(3) Tile is properly laid to drain.

(4) It is not obstructed with sediment or other objects, as disclosed by lamping or rodding.

(5) Discharge is to an area with definite slope away from igloos.

(6) Discharge end of tiles are screened to prevent entrance of rodents.

e. *Replacement of Earth Cover.* Replace the earth cover carefully in two stages using procedures that will protect the waterproofing from damage. Remove rocks from the earth to the extent that it shall contain not more than 15 percent stone or gravel, all passing a one-inch sieve. In the first stage, replace the earth halfway up the height of the igloo and allow to settle for about 14 days, or until noticeable compaction has occurred. At this stage examine the slip sheets for possible breaks at the

*Figure 40. Typical waterproofed igloo.*

earth line. If any are found correct them by splicing and piecing. In the second stage replace the earth to the top of the barrel retaining wall and maintain it at the same height for the length of the igloo. Maintain a slope of approximately 1¾ to 1 free of vertical and horizontal depressions. Maintain sufficient compaction to avoid appreciable slump, slippage, or shrinkage of the finish grade. Bulldozers must not approach the igloo closer than about 3 feet. Finish-grade, hand-rake, and smooth the earth to provide natural drainage (fig. 41). The earth should be provided with vegetation of a type consistent with the landscaping plan for the installation. Remove dirt and other foreign materials from the concrete apron of each igloo and assure that drainage outlets are exposed and opened.

*Figure 41. Finished igloo.*

# SECTION IX
# BRICK AND CONCRETE MASONRY

## 43. General

Use materials, equipment, and methods in masonry construction which conform generally to requirements in appendix III. These specifications cover such masonry units as brick, structural clay tile, and concrete masonry. Stone masonry is not covered in this manual (pars. 61–63, Maintenance and Waterproofing Masonry Walls.)

## 44. Clay and Shale Units

*a. Brick.*

(1) Brick is a commonly used masonry unit, especially in areas where good quality clay or shale are natural deposits. Clay or shale brick is always hardened by baking under high heat. The brick face is either rough or smooth, according to the process used in molding. Those cut from a column of clay mixture by a wire, which leaves their edges turned up, are rough-faced and are known as tapestry or rough-finished brick. Smooth-face bricks are pressed in trays of individual molds.

(2) Generally, common brick is made of a lower quality clay, and is not molded and fired by careful processes as the more expensive face brick. Class H (hard) common brick is used extensively instead of face brick if durable masonry is needed and face-brick appearance is not essential. Class H should be used in places exposed to soil or weather. Class S (soft) common brick is sometimes used for interiors and other places not exposed to soil or weather, but class H or class M (medium) is preferred.

*b. Size of Brick.* The following brick sizes, with variations of plus or minus $\frac{1}{16}$ inch in depth, $\frac{1}{8}$ inch in width, and $\frac{1}{4}$ inch in length, are standard in the United States:

| Type of Brick | Dimensions (inches) | | |
|---|---|---|---|
| | Depth | Width | Length |
| Common | $2\frac{1}{4}$ | $3\frac{3}{4}$ | 8 |
| Rough-face | $2\frac{1}{4}$ | $3\frac{3}{4}$ | 8 |
| Smooth-face | $2\frac{1}{4}$ | $3\frac{7}{8}$ | 8 |

In addition, several special sizes of rectangular brick are made in various sections of the country. Some of these special bricks are made in 2-brick height, 2-brick width, or $1\frac{1}{2}$-brick height or length.

*c. Structural Clay Tile.* Structural clay tile is composed of hollow clay units molded by extrusion and hardened by baking at high temperatures. They are being used more and more extensively as various sizes, shapes, and finishes are designed by manufacturers.

(1) *Sizes.* Faces of hollow tile may vary in size from $2\frac{1}{4}$ by 8 inches (common brick size) to 8 by 16 inches; thicknesses vary from 2 to 12 inches. Two-inch-thick tile, known as soaps, are used as furring tile or fillers. Larger sizes speed up wall construction; one 8- by 16-inch hollow tile requires little more setting time than one brick; yet it occupies approximately the wall space of twelve bricks and their mortar joints. One 5- by 8- by 12-inch hollow tile replaces six bricks.

(2) *Shapes.* Numerous shapes, such as half and quarter sizes, stretcher, header, jamb, sill, cove base, cap, miter, and corner units, are available for various structural needs. Such units simplify laying up a wall.

(3) *Finishes.* Several finishes that produce attractive surfaces and meet different structural needs are available in hollow tile.

(a) Use smooth-finish tile if units are to be exposed or painted.

(b) Use scored finish when plaster is applied on interior walls or stucco on exterior work, (greater expense since two materials and two processes are necessary).

(c) Salt-glaze and ceramic-glaze hollow tiles are self-sufficient structural units with high-grade finish (color if desired). Use where fine appearance is needed. Although this is more expensive than the other two, the wall is completely finished in one process with one material.

## 45. Concrete Masonry

*a. General.* Concrete masonry includes several sizes and types of solid blocks, brick, and hollow building units made of molded and pressed concrete. The concrete is made by mixing portland cement, sand, and water with several types of aggregates such as gravel, crushed stone, cinders, burned clay, or shale.

*b. Quality.* Concrete-masonry units intended for use in walls exposed to weather or soil should be dense and highly resistant to moisture absorption. Units made with lightweight large-size aggregate absorb enough sound to make them suitable for exposed interior wall surfaces in auditoriums, classrooms, corridors, and other places where noise abatement or improved acoustics are required. Units for any given structure should be of the same manufacture, composition, size and appearance, cured by the same process, and free of defects that would interfere with proper installation, strength, durability, or appearance. Hollow load-bearing units should conform to ASTM Designation: C–90. Grade A units should be used in unparged walls below grade, in unpainted walls above grade where subject to frost, and in all load-bearing walls in seismic construction. Hollow non-load-bearing units should conform to ASTM Designation C–129. Solid load-bearing units should conform to ASTM Designation C–145. Masonry units should be delivered for use in an air-dry condition. Test by the procedures outlined in Appendix III. Do not wet masonry units before laying.

*c. Sizes.* Concrete units lay up rapidly because of their large sizes. Special units, such as half and quarter sizes, corner, jamb, joist, bond beam, and header units simplify laying up a wall (fig. 42). (See also National Bureau of Standards Report 3079, pages 30 & 31, figs. 1 & 2.)

## 46. Mortar

*a. Good Mortar.* Mortar generally should comply with ASTM Standards C–270. Good mortar sticks to all masonry surfaces. When properly mixed and used it helps produce weather-resistant, durable masonry of required strength and appearance. Use of fine clean sand, with oversized particles screened out, is important. Oversized particles slow down construction by preventing the mason from shoving units to required position.

*b. Poor Mortar.* Poor mortar containing impurities in the sand, or acids, organic matter, or other harmful substances in the water may mar the appearance of the masonry as follows:
   (1) *Bleeding.* Bleeding means losing water during construction. Water which carries impurities leaves stains when it dries.
   (2) *Efflorescence.* Efflorescence is a white deposit of soluble salts frequently appearing on the surface of masonry walls. Impure water may contain soluble salts which later appear as efflorescence.

   (3) *Weakened units.* Impurities may cause wearing away under extremes of weather action.

*c. Standard Materials.* Experience shows that good results are obtained with mortars of standard materials (cement, sand, lime) which usually are available locally. Cost often determines selection of mortar materials. Paying a premium for special ingredients claimed to increase mortar adhesion and to make mortar more workable, is not justified.

*d. Types.* Characteristics of various mortar ingredients are:
   (1) High cement content produces high strength, quick set, fair workability, some shrinkage in drying, little elasticity.
   (2) High lime content produces low strength, slow setting, easy workability, considerable shrinkage, good elasticity.
   (3) High sand content produces low shrinkage, low strength, poor workability.
   (4) High water content produces easy application, low strength, excessive shrinkage.

*e. Porosity.* Mortar for dense masonry units in exterior walls above grade should be reasonably porous (6 to 12 percent absorption during 24-hour cold immersion) to permit escape of moisture from the walls through the mortar joints.

*f. Recommended Mixes.*
   (1) The following is a partial list of mortar mixtures suitable for average Army building work.

| Type of mortar | Parts by volume | Material |
|---|---|---|
| Lime-cement | 1 | Lime, putty. |
| | 1 | Portland cement. |
| | 6 | Sand. |
| Masonry-cement | 1 | Masonry cement. |
| | 3 | Sand. |
| Slag-cement | 1 | Slag cement. |
| | 1 | Lime putty. |
| | 6 | Sand. |
| High portland-cement | 1 | Portland cement. |
| | 3 | Sand. (10 to 15 percent of cement volume is hydrated lime.) |

*Caution:* **Do not add salts or chemicals to retard freezing. So much is necessary for an appreciable result that mortar strength is seriously reduced. Added salts may cause efflorescence.**

THREE-CORE 8" x 8" x 16"
ALSO 10" & 12" WIDTHS
ALSO HALF UNITS

TWO-CORE 8" x 8" x 16"

8" x 3" OR 4" x 16"
ALSO 9" x 3" OR 4" x 18"

5" x 8" x 12"
ALSO HALF UNITS

3½" x 8" x 12"
ALSO HALF UNITS
(HEIGHT MAY VARY)

TYPES OF CONCRETE-WALL UNITS, STRETCHERS

CORNER UNIT
ALSO HALF UNITS

HEADER UNIT

PIER OR DOUBLE
CORNER UNIT

WOOD SASH
JAMB OR JOIST UNIT
ALSO HALF UNITS

STEEL-SASH
JAMB UNIT
ALSO HALF UNITS

STANDARD SPECIALS FOR 8" UNITS. ALSO MADE IN TWO-CORE TYPE
SIMILAR SPECIALS ARE REGULARLY FURNISHED FOR 10" & 12" UNITS

PARTITION UNIT
ALSO 5⅜" WIDE

WOOD-SASH
JAMB UNIT
ALSO HALF UNITS

STEEL-SASH
JAMB UNIT
ALSO HALF UNITS

PARTITION UNITS

SPECIALS FOR 5" x 8" x 12" UNITS

SPECIALS FOR 3¼" x 8" x 12" UNITS

JAMB UNIT

JAMB UNIT
ALSO 6" WIDE & HALF UNITS

FRACTIONAL UNITS

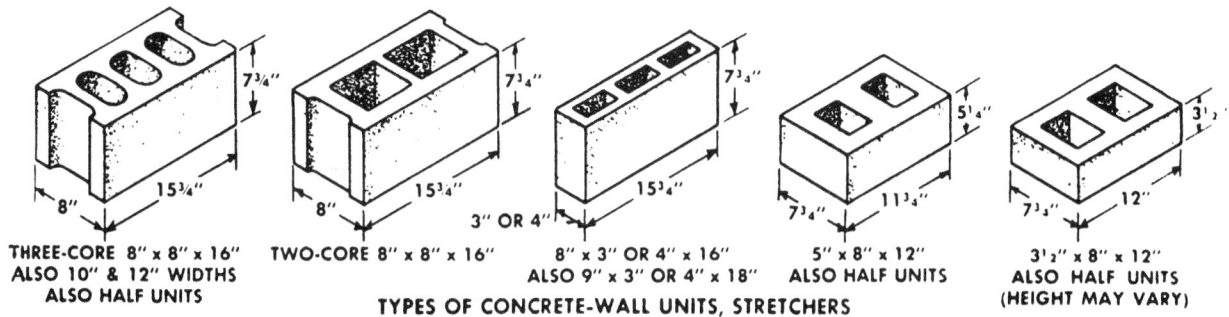

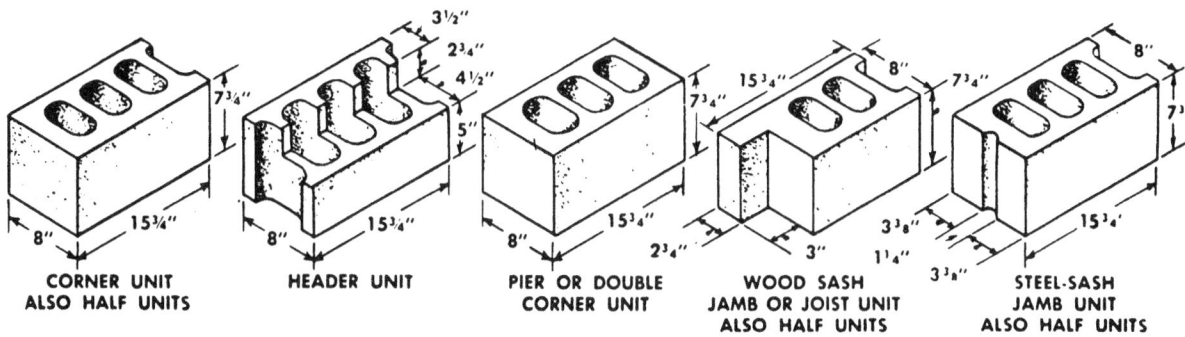

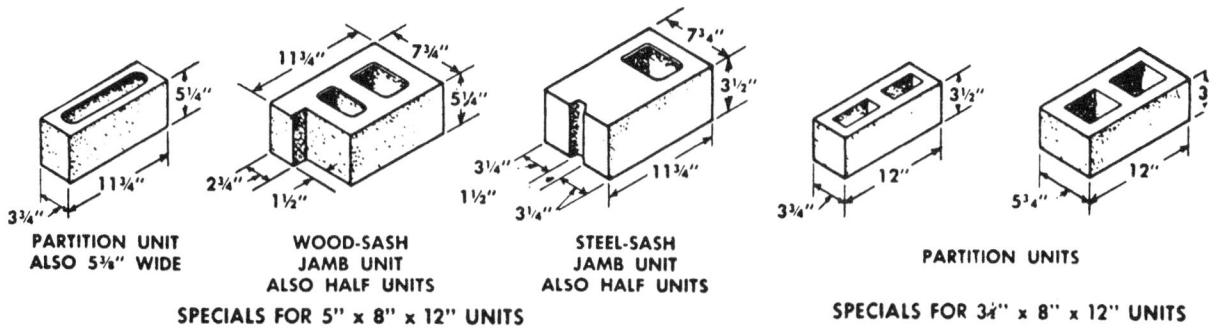

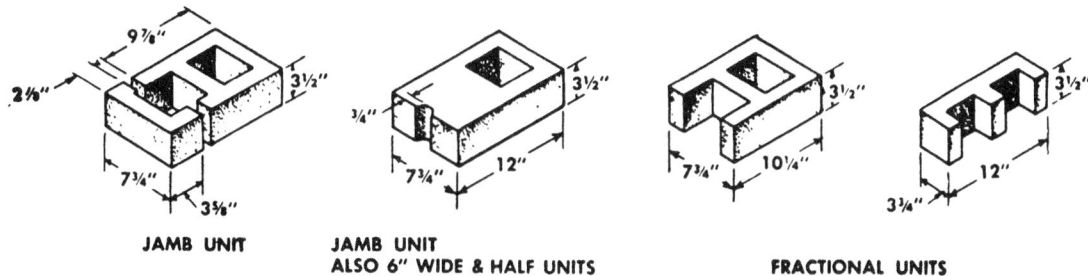

*Figure 42. Typical hollow concrete masonry units.*

(2) Use high early-strength portland-cement for the following masonry construction:
  (a) Piers and walls carrying heavy loads.
  (b) Masonry work under water.
  (c) Fire walls.
  (d) Fireplaces and fireplace flues.
  (e) Free-standing chimneys above roofs.
  (f) Chimney and parapet caps.

g. *Prepared mortar mixes.* Special mortar mixes are sold under various trade names. They are often dry mixes with waterproofing admixtures. Their chief advantage is the ease of getting a proper mixture. They may be purchased under Federal Specifications SS–C–181 Cement; Masonry.

h. *Lime.*
  (1) *Types.* The following limes can be purchased for structural purposes:
    (a) *Hydrated lime.* Types M, masons, and F, finishing, conforming to Federal Specification SS–L–351, Lime; hydrated (for) structural purposes.
    (b) *Quicklime.* Types C, calcium, and M, magnesium, in lump, granular, pebble, ground, or pulverized form, conforming to Federal Specification SS–Q–351, quicklime; for structural purposes.
  (2) *Form.* Lime used in the form of well-aged putty is preferred.
  (3) *Hydrated-lime putty.* Although dry lime is sometimes added to mortar mix without preliminary soaking, plasticity of hydrated-lime putty is improved by soaking the hydrate for at least 12 hours. Sift dry lime into water so every particle is thoroughly wetted.
  (4) *Slaking quicklime.* Dry quicklime cannot be used for structural purposes until it is slaked. The method of slaking largely determines quality of the finished putty. Different brands of quicklime vary considerably in their action when water is added to them. Weather conditions have a decided influence.
    (a) *Sample slake.* To determine characteristics of a particular lime, put about 1 quart of lime in a bucket, add water to cover, and note how long it takes for slaking to begin. Slaking has begun when pieces split off the lumps, when lumps or pebbles crumble, or when ground or pulverized lime shows action in the water. If slaking begins in less than 5 minutes, lime is quick-slaking;

from 5 to 30 minutes, medium-slaking; over 30 minutes, slow slaking. Use water at same temperature for both test and field slakes.
  (b) *Slaking procedure.*
    1. For quick-slaking lime, always add lime to water, not water to lime. See that lime is covered with water. Watch lime closely, and at first appearance of escaping steam, hoe thoroughly and quickly, adding plenty of water to stop steaming. Have plenty of water available for immediate use. If possible, use a hose throwing a good stream.
    2. For medium-slaking lime, place lime in mixing box and add water to half-submerge the lime. Hoe occasionally if steam starts to escape. Add a little water occasionally if necessary to prevent putty from becoming dry and crumbly. Be careful not to add more water than necessary nor too much at a time.
    3. For slow-slaking lime, place lime in mixing box and add water to moisten it thoroughly. Let it stand until action starts, then add water, a little at a time, taking care that mass is not cooled by fresh water. Do not hoe until slaking is practically complete. If the weather is very cold, hot water is preferred. If hot water is not available, cover mixing box to retain heat.
  (5) *Preparation of mason's mortar.* After all slaking and soaking has ceased, add part or all of the sand required and store before use for at least 24 hours, until putty has completely cooled. Mortar that has stiffened because of chemical reaction (hydration) should not be used. Except as provided below, mortar should be used and placed in final position within 2½ hours after mixing when the air temperature is 80° F. or higher, and within 3½ hours after mixing when the air temperature is less than 80° F. Discard mortar not used within these time intervals. When the cement or cements have been tested and initial set has been determined in accordance with ASTM Designation C–266, the time during which mortar should be used

may be established at 1 hour less than initial set when the air temperature is 80° F. or more, or ½ hour less than initial set when the air temperature is less than 80° F. Mortars that stiffen within the allowable time intervals for use may be retempered to restore workability by adding water as needed.

## 47. Second-Hand Masonry Units

Brick and other burned-clay units are durable and can be salvaged from old buildings for reuse. If bricks or other units are used second-hand, sort according to quality. Chip off mortar and wire-brush the units. Disadvantages in using second-hand units include:

*a.* Lack of uniform quality because under-burned or otherwise poor units originally used as backup or in partitions may have been mixed with well-burned exterior units.

*b.* Presence of old mortar due to incomplete cleaning hinders good construction.

*c.* Installing bricks which were subjected to chemicals or vapors thus resulting in discoloration or efflorescence.

## 48. Workmanship

Care in masonry construction reduces the subsequent need for maintenance and repair. Competent supervision of construction work is necessary, especially when work is done by unskilled or inexperienced labor. Do not sacrifice good workmanship for speed. Lay masonry units on level beds and to plumb vertical lines. Quality of materials affects workmanship. Highly plastic mortar works easily, improves workmanship, and increases speed of laying.

## 49. Freezing Weather

Although laying brick in freezing weather is not recommended, particularly if the brick must be wet, masonry structures can be built in freezing weather. When working in freezing weather is necessary, heat all materials so they will remain above 32° F. until they have been placed and suitably protected. Protect finished masonry from freezing for at least 48 hours after laying.

## 50. Warm Weather

Clay masonry units laid dry in warm weather absorb a large amount of water from the mortar, causing excessive mortar shrinkage and resultant loss of bond. This loss of bond soon appears as a hairline crack between unit and mortar. Make sure brick and similar masonry units are damp when laid during temperatures of 40° F. and above. Wet units thoroughly at least 1 hour before laying. In late afternoon thoroughly wet brick and similar units to be used during the first 2 hours of the following day's work. Permit no surface water on units at time of laying.

*Caution:* **Since masons generally do not like to use bricks which have been properly wetted because wear on fingers is greater, check continuously to make sure units are laid damp. Concrete masonry units, however, should always be laid dry.**

## 51. Bonds

See figure 43 for typical brick-bond patterns and figure 44 for typical concrete-masonry unit bonds. Bonds can be varied or emphasized by using masonry units of different colors and textures and by changing the type of exposed mortar joints.

## 52. Joints

*a. General.* The quality of interior and exterior mortar joints is important in producing a masonry wall of proper strength and resistance to water. Ideal masonry construction includes smoothly spread mortar beds, end surfaces of units buttered with enough mortar to fill the joint completely, vertical back joints completely filled by slushing, and weathered or concave exterior face joints. In building brick walls, shove all backing brick into place after first placing a heavy bed of mortar. When hollow units are used, completely mortar-coat all bearing surfaces.

*b. Width of Joints.* Joints approximately ⅜ inch wide are generally best for forming brick bonds and patterns, since the width of two headers plus this joint width equal the length of a standard stretcher. Concrete masonry units are usually laid with ¼ inch to ⅜ inch joints. Color texture of joint mortar affects appearance and quality of the finished wall and should be kept uniform.

*c. Type of Joints.* Seven types of exposed face joints are shown in figure 45. All can be used with solid units, but raked and stripped joints are not recommended for hollow units.

    (1) *Struck joints.* A struck joint is widely used on interior walls. It is easy to make because it can be struck below eye level and on the wall surface opposite the scaffold. It is not suitable for exterior

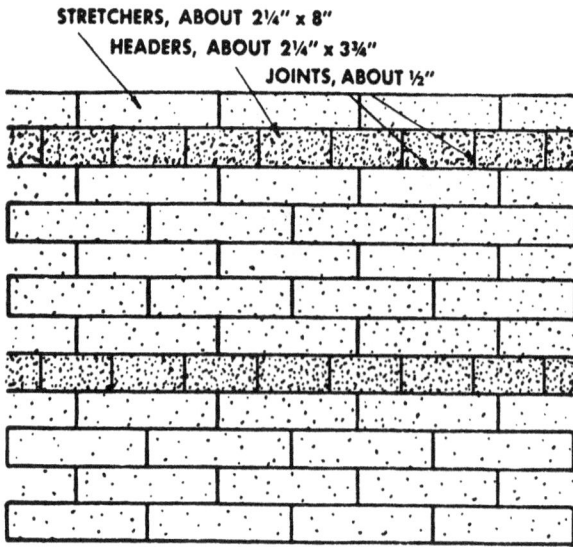

STRETCHERS, ABOUT 2¼" x 8"
HEADERS, ABOUT 2¼" x 3¾"
JOINTS, ABOUT ½"

COMMON

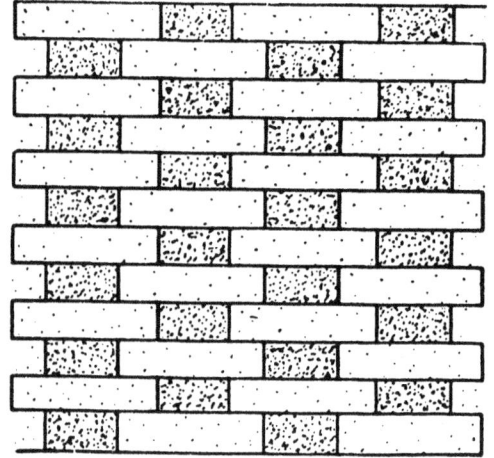

FLEMISH
(ALTERNATE HEADERS AND STRETCHERS)

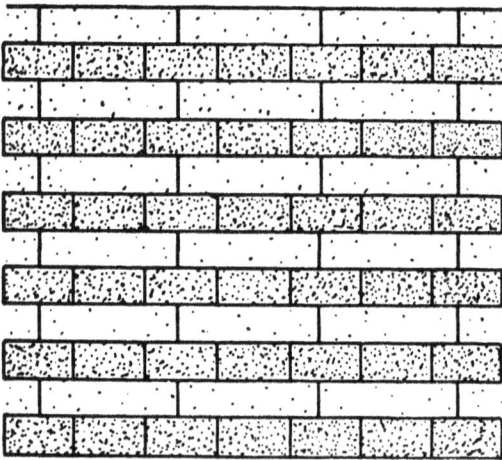

ENGLISH
(ALTERNATE COURSES OF HEADERS AND STRETCHERS)

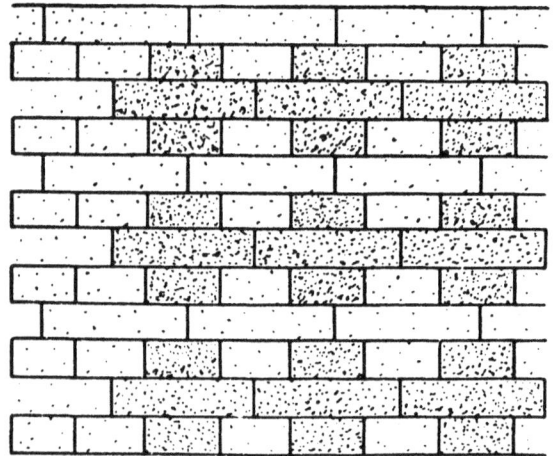

CROSS OR DUTCH

*Figure 43. Typical brick bonds.*

AGO 10179A

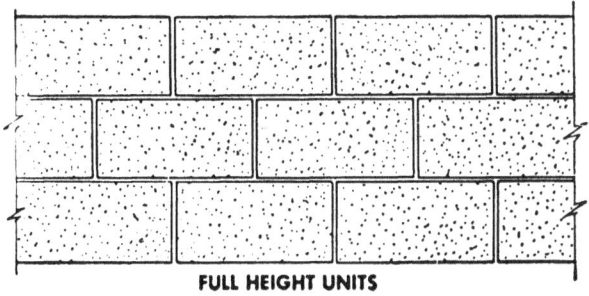

**FULL HEIGHT UNITS**

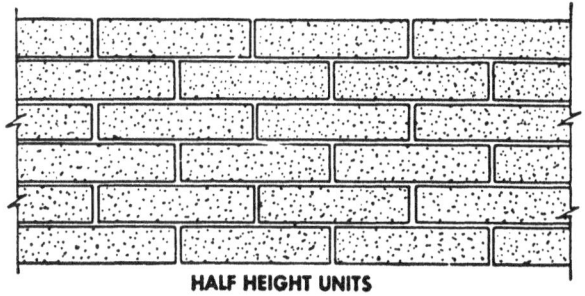

**HALF HEIGHT UNITS**

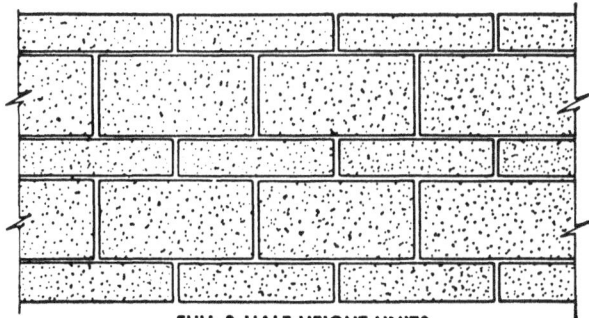

**FULL & HALF HEIGHT UNITS**

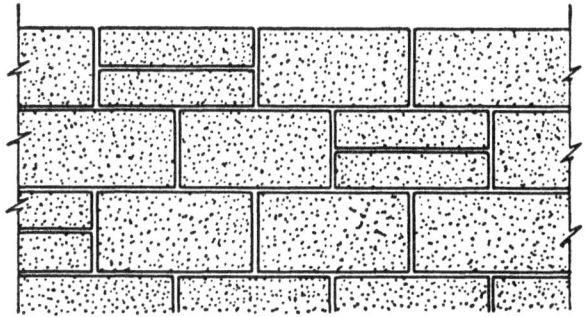

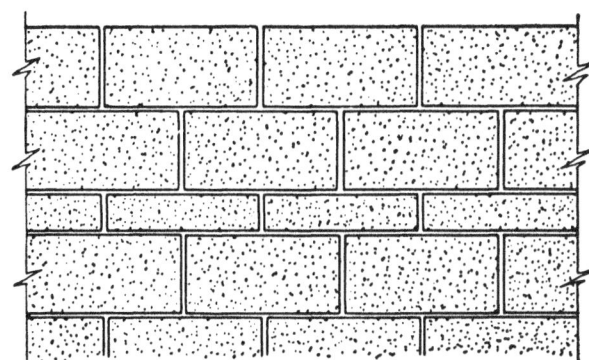

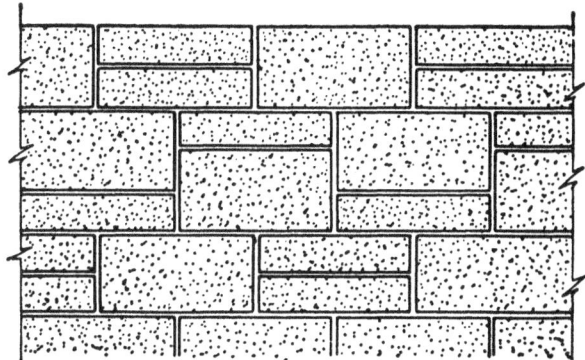

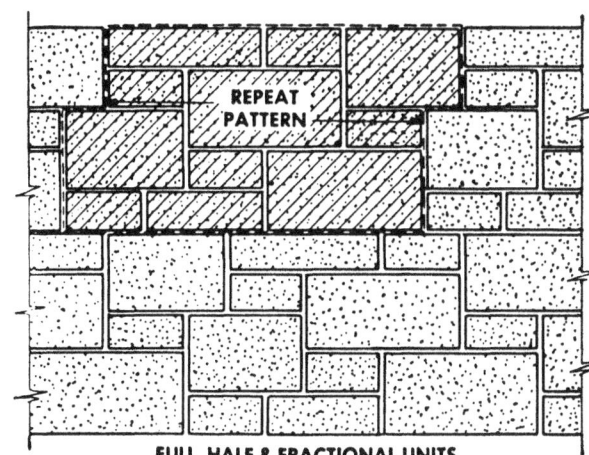

**FULL & HALF HEIGHT UNITS**

**FULL & HALF HEIGHT UNITS**

**FULL & HALF HEIGHT UNITS**

REPEAT PATTERN

REPEAT PATTERN

**FULL, HALF, & FRACTIONAL UNITS**

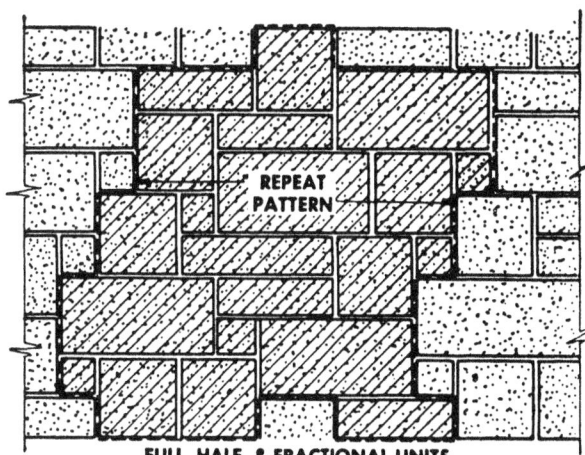

**FULL, HALF, & FRACTIONAL UNITS**

NOTE:
ABOVE PATTERNS CAN BE PRODUCED WITH STANDARD UNITS—8" x 16" FACE,
5" x 12" FACE, AND 3¾" x 12" FACE IN USUAL WALL THICKNESSES.
MANY OTHER PATTERNS MAY BE WORKED OUT BY THE ARCHITECT.

*Figure 44.   Typical concrete-masonry-unit bonds.*

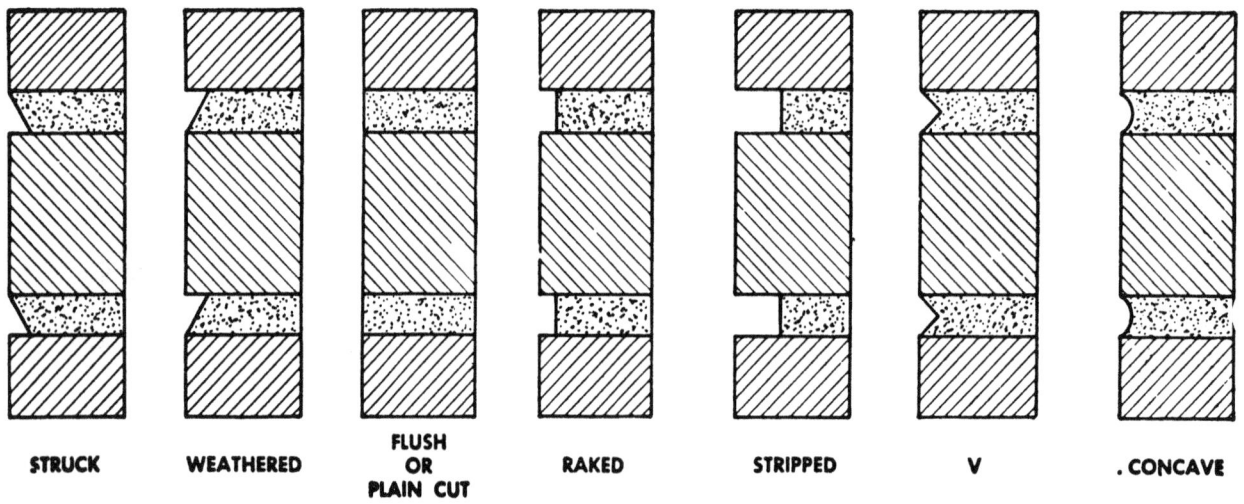

STRUCK      WEATHERED      FLUSH OR PLAIN CUT      RAKED      STRIPPED      V      . CONCAVE

*Figure 45. Typical mortar joints for brick and concrete masonry.*

walls because the small horizontal shelf permits water to enter the wall, particularly if the mortar shrinks.

(2) *Weathered joint.* Make weathered joints by forming flush or plain-cut joints, then finishing with a trowel after mortar sets thumb-print hard. An efficient, water-shedding low-cost joint, is preferable to a struck joint because the upper brick forms a drip and no shelf is left on the lower brick.

(3) *Flush or plain-cut joint.* Make flush or plain-cut joints by cutting mortar flush with wall surface. If a rough texture is desired, do not touch mortar with the trowel after the first cut. This joint does not resist water entrance as well as a tooled joint.

(4) *Raked joint.* Form raked joints by allowing mortar to stiffen slightly and then raking joint about ½ inch deep with a special tool.

(5) *Stripped joint.* Make stripped joints by inserting wood strips about 1 inch deep in the face of the joint. After the mortar sets enough to hold brick, remove the wood strips. Both raked and stripped joints throw horizontal shadow lines, giving the wall an interesting appearance. Since they do not resist moisture as well as tooled or weathered joints, they are not recommended for extensive use.

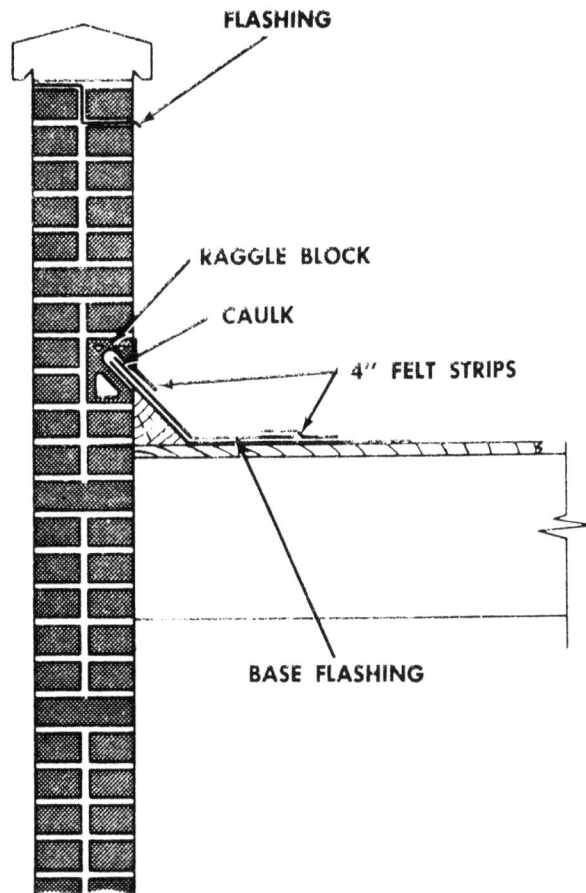

*Figure 46. Flashing at intersection of roof and wall.*

(6) **V**-*joints and concave joints.* Make V-joints and concave joints (sometimes called tooled joints) by forming flush or plain-cut joints, then finish with special tools (V-shaped or round bent rod) after mortar sets thumbprint hard. These joints are easily formed, inexpensive, and weather resistant.

## 53. Flashing

a. *General.*

(1) Flashing is the impervious membrane placed at certain places in masonry for the purpose of excluding water or for collecting any moisture that does penetrate the masonry and direct it to the outside wall. Flashings are installed at the heads and sills of window openings, and, in some buildings, at the intersection of the wall and roof. Where chimneys pass through the roof, the flashing should extend entirely through the chimney wall and turn up for a distance of 1 inch against the flue lining. Such flashing is embossed to provide a bond.

(2) The edges of the flashing are turned up as shown in figure 47 to prevent drainage into the wall. Flashing is always installed in mortar joints. Drainage for the wall above the flashing is provided by placing ¼-inch cotton-rope drainage wicks in the mortar joint just above the flashing membrane at 18-inch spacings. Drainage may also be accomplished by holes left after dowels placed in the proper mortar joint are removed.

(3) Flashing required at the intersection of the roof and wall (fig. 46) is always installed to prevent leakage between the roof and the wall.

b. *Flashing Materials.* Copper, lead, aluminum, galvanized iron, and bituminous roofing paper are materials that may be used for the flashing membrane. Copper is generally preferred but it will stain the masonry when it weathers. If this staining is undesirable, lead-coated copper should be used. Bituminous roofing papers are cheapest, but not as durable and will have to be replaced periodically in permanent construction. The cost of replacement is many times the cost of installing high-quality flashing. Corrugated copper flashing sheets having interlocking watertight joints at points of overlap and producing a good bond with the mortar, are available.

c. *Installation of Flashing.*

(1) In placing flashing, a ½-inch-thick bed of mortar is spread on top of the brick and the flashing sheet pushed firmly down into the mortar. The brick or sill that goes on top of the flashing is forced into a ½-inch-thick mortar bed spread on the flashing.

(2) Details for the proper installation of flashing at the sill of a window are shown in figure 47.

## 54. Annual Inspection of Masonry Walls

Inspect masonry walls at least annually, and oftener as local conditions or deterioration indicate. During the inspection, give careful attention to the condition of the mortar, detection of sanding, separation of mortar from the units, adequacy of caulking around window frames and door frames, (pars. 38 and 55f), condition of roof flashing, joints of parapet units, and porosity of parapet wall units. Many leaks attributed to roof trouble are caused by open joints between parapet stones or by mortar separation in vertical joints in masonry. Porosity of units themselves is not a common cause of leakage.

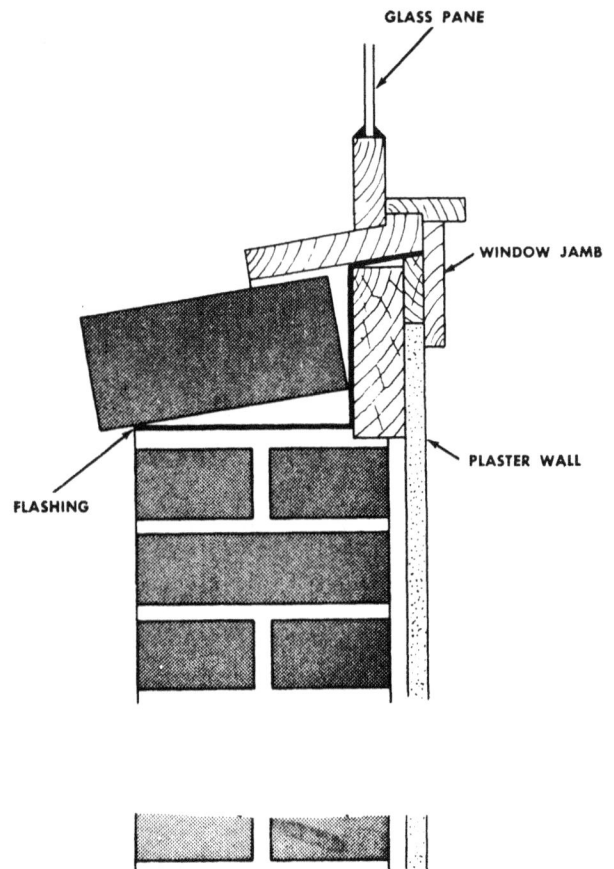

*Figure 47. Flashing at window sill.*

# SECTION X
# GLASS BLOCK MASONRY

## 55. Materials

a. *Glass Blocks.*

(1) *General.* Glass blocks are hollow, partially vacuumized units with lipped edges. They are formed of clear, colorless glass by fusing two halves together at high temperature. Glass blocks are not load-bearing materials; they should never be subjected to imposed loads other than their own weight. They are particularly suited for both interior and exterior walls of such buildings as offices, shops, schools, and hospitals where a combination of light and privacy is essential. They are durable and need practically no maintenance other than occasional cleaning. Ease of cleaning is of further advantage in hospitals, kitchens, food-storage rooms, and other places with high standards of cleanliness.

(2) *Design.* Glass blocks are made in a wide variety of designs. Three general types of blocks, functional, decorative, and all-purpose, are used. The functional type controls the direction in which light rays pass through the glass. The general-purpose type diffuses light without controlling its direction. The decorative type is intended for use where design is the principal consideration.

(3) *Surfaces.* Mortar-bearing surfaces of the blocks may be smooth or corrugated. They should be coated at the factory with a grit-bearing material resistant to water and alkali, to form a bond for mortar.

b. *Mortar.*

(1) *Materials.* The materials used in mortar for glass block masonry are:

(a) *Cement.* Conforming to Federal Specifications SS-C-192, cement; portland, type I.

(b) *Lime.* Conforming to Federal Specification SS-L-351, lime; hydrated, or to Federal Specification SS-Q-351, quicklime. A cubic foot contains 40 pounds of hydrated lime or 25 pounds of quicklime. Soak hydrated lime at least 12 hours; slake quicklime and store at least 24 hours or until completely cooled; screen both before using.

(c) *Sand.* Conforming to Federal Specification SS-A-281, aggregate; (for) portland cement concrete, Class 1. Not more than 12 percent by weight should pass a No. 100 sieve and 100 percent should pass a No. 8 sieve.

(d) *Water.* Clean, fresh, and free from silt, salts, or other foreign matter.

(2) *Mixing.* Use the following proportions: one part portland cement, one part lime putty, and four parts sand. Mix mortar to a stiff consistency, but not too stiff to be workable. Mortar for glass block masonry should be drier than for brick masonry.

c. *Asphalt Emulsion.* Use a clay-base asphalt emulsion suspended in water with a consistency which provides a heavy, even coating free of lumps and voids.

d. *Expansion-Joint Filler Strips.* For expansion joints, use strips about 4 inches wide, 25 inches long, and ½ inch thick, of fibrous glass or similar material bound between two layers of asphalted felt.

e. *Oakum.* Use a loose-fiber nonstaining type of oakum treated to prevent dry rot.

f. *Calking Compound.* Use calking compound of knife or gun consistency. It should be a nonhardening polysulfide sealent complying with Federal Specification TT-S-00227.

g. *Reinforcement and Anchors.*

(1) *Reinforcement.* Reinforcement is necessary in glass block panels for stability of the panel; it will not make it capable of carrying an imposed load. Various metal strips are manufactured for reinforcement. One type is a 20 gage galvanized metal strip 1¾ inches wide with rectangular slots cut by a die. Another is 2 inches wide double wire mesh, formed by two longitudinal wires about 0.150 inch in diameter crossed by wires 0.105 inch in diameter electrically welded in place at regular intervals.

(2) *Panel anchors.* Anchors are required to fasten panels to sides of openings. This is accomplished by use of 20-gage galvanized metal strip wall ties 1¾ inches wide and 24 inches long imbedded in horizontal mortar

joints and anchored to or imbedded into sides of frame. A ½-inch crimp or offset is required in panel anchors of jambs to permit expansion of panel.

## 56. Laying Glass Blocks

Lay glass blocks in checkerboard patterns unless otherwise directed. Fill all horizontal and vertical cracks with mortar. Do not furrow mortar bed joints or let them bridge expansion joints. Lay blocks plumb and true, with courses level and visible mortar joints not less than ¼ inch or more than ⅜ inch wide. Before the mortar attains its initial set and while it is still plastic, rake out joints to expose clean, sharp edges of blocks and give joints a smooth concave tooling immediately afterward. Clean all surplus mortar from the panel as joints are tooled.

*a. Panel Reinforcing.* Reinforcing ties must be continuous from one end of the panel to the other. If more than one piece is necessary in this length, lap ends of meeting ties at least 6 inches. Bed ties completely in mortar. Do not let ties bridge expansion joints or extend beyond edges of glass panels. Place wall ties in horizontal mortar joints of panels at the following intervals:

(1) For 5¾-inch blocks, every fourth course.
(2) For 7¾-inch blocks, every third course.
(3) For 11¾-inch blocks, every course.

*b. Wall Anchors.* Build wall anchors in the jambs at the same courses with panel reinforcing. Bed one end of each wall anchor at least 12 inches in wall, crimp or offset anchor within expansion joint, and embed other end in mortar joint of glass-block panel.

*c. Sills.* Give sills a heavy coat of asphalt emulsion and allow it to dry at least 2 hours before spreading mortar.

## 57. Expansion Joints

Provide expansion joints not less than ½ inch wide at jambs and head of each glass-block panel. Do not fill these joints with mortar. Make the joint between top of the glass-block panel and bottom of the lintel deep enough to compensate for any lintel deflection. Provide intermediate expansion joints where glass-block panels are over 20 feet in width or length. Pack expansion joints either with oakum or premolded expansion-joint filler strips. Secure expansion-joint packing and strips to jambs and head with asphalt emulsion. Calk interior and exterior perimeters of glass-block panels with compound to depth of at least ½ inch.

## 58. Cleaning

Wash off all mortar splashes and other foreign matter on completion of glass-block work, after mortar attains its final set. Be careful not to damage mortar joints or injure mastic calking in waterproofed expansion joints. Do not use strong alkaline solutions, coarse scouring powders, steel wool, or wire brushes; use any neutral cleanser or mild soap solution. Rinse surfaces with clear water, wipe with a clean dry cloth, and leave the glass-block panels clean and bright. Use the same method for regular maintenance cleaning. Keep glass blocks clean to transmit light effectively.

## 59. Maintenance and Repair

*a. General.* Glass-block panels require little maintenance other than occasional cleaning and periodic inspection of joints. They can be damaged seriously either by superimposed loads transmitted through excessive deflection in beams and lintels, uneven settlement of foundations and footing, or by impact. Correct causes of damage before repairs are made.

*b. Repair of Panels.* To repair panels, remove cracked or broken glass blocks. Chip off fragments of broken glass adhering to undamaged blocks, taking care not to injure wall ties and anchors. Clean old mortar from exposed panel reinforcing, wall anchors, and mortar-bearing edges of adjoining blocks. Install new block plumb and true with joints matching those of adjacent block, as described in paragraph 56 of this section. Clean out ineffective calking compound and recalk as outlined in paragraph 55f.

*c. Replacement of Panels.* Replace panels, using materials and methods described herein. Glass blocks, wall ties and anchors, expansion-joint filler strips, and calking compounds, should match existing work as closely as possible.

*d. Inspection.* Inspect glass-block walls at least once a year and after each exposure to impact damage.

# SECTION XI
# MAINTENANCE AND WATERPROOFING OF MASONRY WALLS

(See also Section VII, Maintenance and Waterproofing of Concrete Walls.)

## 60. Efflorescence

Efflorescence on masonry always indicate trouble. It usually appears as a light powder or crystallization caused by water-soluble salts deposited on the surface when water evaporates within the mortar or the masonry unit. Aside from its unsightly appearance, it is evident that enough moisture may be penetrating the wall to cause disintegration of the masonry.

*a. Conditions Producing Efflorescence.* The two conditions which generally produce efflorescence are presence of water-soluble salts in masonry units, mortar, or both, and moisture to carry salts to the wall surface. This latter may be excess moisture in mortar, moisture caused by extra soaking of units before laying, or moisture taken up during storms after erection of the wall.

(1) *Water-soluble salts.* Soluble salts may be present in brick, hollow tile, concrete blocks, or mortar. Tests have shown that only a small percentage (probably not more than 10 percent) of well-burned clay, sand-lime brick, and hollow-clay-tile contribute to efflorescence. Second-hand brick, because of its uncertain origin and previous contact with mortar and plaster of unknown composition, may effloresce. Concrete blocks are often made of materials containing efflorescing salts. Portland cements, limes, and sands used in mortars often contain soluble salts that cause efflorescence. The wick test (app. I) is recommended for determining presence of soluble salts in masonry units and mortar ingredients.

(2) *Excessive moisture.* Since moisture is necessary to carry soluble salts to exterior masonry surfaces, efflorescence is evidence that construction faults have allowed moisture to enter the wall. These problems are covered in paragraph 37 above.

*b. Determination of Source.* The source of efflorescence on walls can be found if close inspection is made when it first appears. If it appears at the edges and not near the center of the masonry unit, the mortar probably is at fault. Efflorescence at the center of the unit and not near the edges indicates that the masonry unit is the source. If efflorescence covers the whole wall, both unit and mortar may be responsible.

*c. Repointing.* Repointing effectively corrects defective mortar joints in a masonry wall. Successful results require the best materials and skilled, painstaking labor under constant, competent supervision.

*d.* (1) *Removing old mortar.* Since removing mortar by hand with hammer and chisel is difficult and expensive, most masons who specialize in repointing use properly guarded portable electric-driven carborundum grinding wheels to simplify the job. Proceed as follows:

(a) Chip out all loose mortar. Remove all old pointing mortar if much of it is in poor condition, so new pointing is continuous and immediate additional repair work will not be needed. Cut out mortar at least ¾ inch deep to provide a bed for new mortar.

(b) Clean out opened joint by brushing or airblasting. Wet all surfaces of open joint just before applying new pointing mortar because dry surfaces take water needed by mortar for proper hardening.

(2) *Recommended mix.* Mix mortar rather dry (about the consistency of putty) in small batches. The following mix is recommended for repointing; one part portland cement, white or natural; two parts fine clean sand, white or colored, screened free from pebbles.

(3) *Application.* Push new mortar back into cleaned joint in a continuous forward operation in one direction from starting point. This expels possible air pockets and forms a firm contact with old mortar at back of joint. Cut mortar flush with wall face, allow it to become thumbprint hard, and use a suitable tool to give desired joint finish. In the finished operation apply considerable pressure to force mortar back

into joint and to form contacts with masonry surfaces.

(4) *Protection.* Protect all fresh repointing from direct exposure to hot sun and drying winds until it has set hard.

## 61. Waterproofing and Dampproofing Masonry Walls
(par. 62*p*)

This paragraph is also applicable to concrete walls to the extent discussed. So far as possible, keep water from accumulating under buildings and around foundations and basement walls. Divert water away from buildings by proper grading under buildings and at least 10 feet from them. Connect downspouts to storm sewers (not sanitary sewers) or ditches leading away from buildings. If water passing through a wall is not under pressure, it can be stopped by mopping, troweling, spraying, or brushing one of several liquid or semiliquid waterproofing materials on the inner surface of the wall. If the water is under some pressure, this method is not dependable and the mopped-on, multiple-membrane system (2 or more layers) of waterproofing is recommended. Open-joint or perforated drain tiles set outside the wall leading to dry walls or open ditches are recommended for reducing water pressure on exterior wall surfaces.

*a. Walls Below Grade.* Waterproof basement walls which extend down into wet soil. Applying waterproofing on exterior surface during construction, before placing the backfill, is less expensive and more effective. Make the waterproof coating continuous from footings to finished grade lines, using the following methods:

(1) *Membrane.* The mopped-on multiple-membrane method is the only dependable way to keep out water under considerable pressure. The built-up membrane can also adjust itself to wall settlement, without breaking. For application see appendixes IV and V. Membrane sandwiched in between layers of walls and floors give best results (fig. 48).

(2) *Bituminous mastic.* Two coats of bituminous mastic on exterior concrete surfaces ordinarily provide satisfactory dampproofing. This mastic may be an asphalt or coal-tar pitch with or without an added solvent plus an inert filler such as silica sand or asbestos fiber (app. V).

(3) *Cement plaster.* An inexpensive method of limited waterproofing below-grade walls is to plaster exterior surfaces with cement mortar. This mortar may be plain, or water resistance can be increased by adding calcium stearate, ammonium stearate, or other equivalent water repellent material in amounts equal to about 3 percent of the cement weight, (par. 19*d*(5) and par. 62*m* and *n*). Although the cement-plaster method is efficient, its rigidity is an undesirable feature. Structural defects which crack the wall may also crack the plastering. This method may be adequate, however, if the wall is subjected only to occasional dampness and not to continuous water pressure. Portland cement plaster finish applied over masonry walls has good bond and requires little maintenance. Strip the loose plaster from the masonry areas. Rake out cracks a half inch in width, the full depth of the plaster. Before applying new material thoroughly wet down area, and drain. Patch with a sand-cement mixture of approximately one part portland cement, to which 10 percent lime and 2½ parts sand have been added. The mixture should be made to match the existing finish as nearly as possible when dry.

(4) *Metallic (iron) powder.* The metallic method of waterproofing consists of applying a

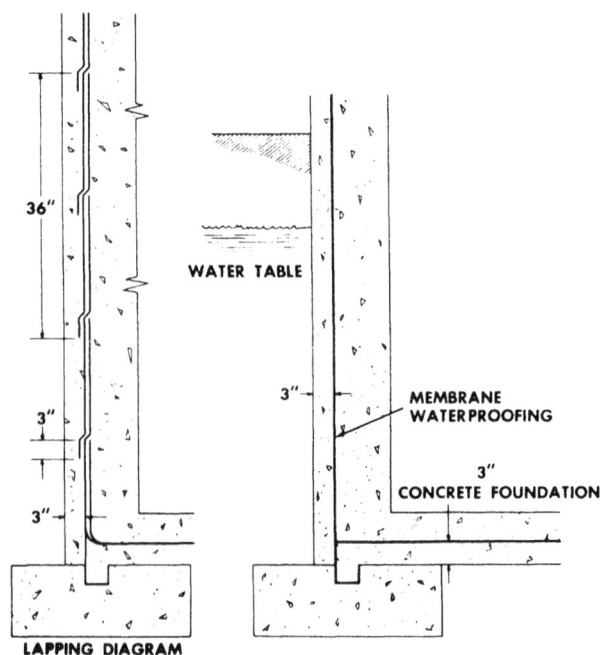

*Figure 48. Typical membrane waterproofing.*

mixture of metallic (iron) powder, cement, sand and water over leaking walls and floors of poured concrete, or over masonry walls of concrete blocks, hollow tiles, cinder blocks, or porous brick. This method cannot be used on glazed or other non-absorbent surfaces.

(a) *Materials.* Use the following materials:

1. *Metallic powder* (par. 62e). Exceedingly fine particles of iron refined relatively free of oil, grease, and nonferrous metals, processed to oxidize (rust) quickly without forming gas bubbles.

2. *Portland cement.* Federal Specification SS-C-192b, cement; portland, type I.

3. *Sand.* Sand should be clean, hard plastering sand; free from loam, silt or other impurities; and well graded from fine to coarse (TM 5-621).

4. *Water.* Water should be potable.

(b) *Action of metallic waterproofer.* When metallic powder is mixed with water and applied as a brush or slush coat; or mixed with cement, sand, and water and applied as a pointing mortar or plaster coat—it oxidizes and expands, filling voids left by evaporating water. This treatment is suitable for interior or exterior application and resists considerable water pressure when done thoroughly. Do work in accordance with paragraph 62 or as recommended by manufacturer of the metallic aggregate used.

b. *Walls Above Grade.* Poor workmanship is the chief cause of leaks in masonry walls above grade. Unless masonry units are laid in full beds of good mortar with all joints completely filled walls are likely to show leaks soon after the first rainy season begins. Hairline cracks between mortar and masonry are definite points of leakage.

(1) *Transparent waterproofers.* Using transparent waterproofing materials to repair leaks in walls above grade is justified only when defects in construction have been made good, and when it is evident that moisture causing efflorescence is entering the vertical wall face because of porous materials; not defective joints (pars. 52 and 62). Select the type of transparent waterproofing material which experience shows is most effective under local climatic conditions. Do not depend on results of laboratory experiment; none of these water-

proofers is effective after 2 or 3 years, except to keep the wall clean and preserve the natural appearance.

(a) *Materials.* Transparent compounds for waterproofing masonry walls which are somewhat effective are:

1. Paraffin or very heavy mineral oils in solution with light mineral spirits.

2. Metallic soaps of aluminum, zinc, and the like, and salts of fatty acids.

3. Varnishes, usually mixtures of organic oils and gums.

4. Sodium silicate, fluosilicates, and similar materials in water solution.

5. Colorless silicone resin base water-repellent, Federal Specification SS-W-00110. (Lasts 2 to 3 years.)

(b) *Preparation of surfaces.* Fill all holes and cracks in the wall face before applying waterproofing materials. Transparent surface-applied material will not fill holes in masonry or correct structural defects.

(c) *Application.* The solutions are intended to penetrate the masonry pores and are generally applied cold, however, penetration of both sodium silicate and fluosilicate is considerably increased by heating to 170° F.

(2) *Portland-cement paint.* Portland-cement paint hides the surfaces and helps waterproof by filling surface voids and crevices. It is intended for use on porous surfaces of concrete, cement stucco, common brick, concrete and cinder blocks, and similar materials. It bonds with porous surfaces to form a highly durable concrete finish with good waterproofing qualities. The principal ingredients of this paint are: White portland cement containing a water repellent, such as 1 percent to 3 percent calcium stearate; a material to hasten the set, such as 3 percent to 5 percent calcium chloride; a filler such as lime; mineral color pigments; and water. Also, very fine sand or other siliceous materials are added to Class B paint to fill voids in open texture concrete or masonry surfaces. Savings can be made by purchasing the ingredients (less than 50c per gal.) and mixing paint on the site. This practice also offers more flexibility. If paint is prepared, measure all quantities of materials except water in order to prepare

additional batches of like colors. Some color pigments, such as greens, fade more readily than others. Wet film cures out much lighter in color when dry. Avoid the use of multicolors and deep shades.

(a) *Materials.* The following are in general use:

1. Portland-cement paint conforming to Federal Specification TT–P–21, paint; cement-water, powder, is available in two types and two classes of each type:

   Type I, 65 percent portland cement

   Type II, 80 percent portland cement

   Class A, without siliceous aggregate, for general use

   Class B, with siliceous aggregate, for open-textured walls.

2. SS–C–192 cement, portland-white.

3. Coloring mineral pigments finely ground.

4. O–C–106 Calcium chloride; hydrated, technical grade.

5. SS–L–351 Lime; hydrated for structural purposes.

(b) *Preparation of surfaces.*

1. Do not apply portland-cement paints until the masonry wall is at least 30 days old. Allow poured concrete walls to dry for several months if possible before applying these paints. Delaying the painting allows weather to disperse the form oil, enables surface suction to become more nearly uniform, permits appearance and removal of efflorescence before painting, covers checking which develops in masonry surfaces, and causes less crazing of the paint film. Removal of efflorescence is important when dark paint shades are used because efflorescence is more noticeable on dark than on light surfaces.

2. Make sure surfaces to which portland-cement paint is applied are clean and free of loose dirt, scale, oil, form lacquer, or other substances which prevent paint from striking in and bonding to the wall. Portland-cement paint is not satisfactory when applied over paint films other than cement water paint.

3. Make surfaces rough or porous enough to hold the paint. Concrete cast against plywood or metal forms and some types of masonry units are sometimes so smooth that good paint adhesion cannot be obtained. Roughen such smooth surfaces by light sandblasting, dry rubbing with No. 16 carborundum blocks, or washing with 20 to 30 percent builder's acid (muriatic). Rinse thoroughly with water after the acid wash.

4. Repair all cracks and inspect and repoint all deficient mortar joints.

(c) *Dampening the wall.* Before applying paint, wet the wall thoroughly with a garden hose giving a fine spray. This water spray controls surface suction and provides additional moisture which assists proper hardening of the paint. Dampening the wall by dashing water with a brush is not adequate. Apply spray so as to maintain a continuous dripping wet surface from the time spraying is begun until painting in each small area is ready to start. This spray period should be a minimum of one hour depending upon the ambient humidity and type of wall. The wet surface should be permitted to dry sufficiently before painting to permit all beads of moisture to disappear, usually 5 to 10 minutes.

(d) *Preparation of paint.*

1. When preparing cement-water paint, follow these instructions closely. First reduce the dry material to a very stiff paste by adding water in small portions, stirring constantly. (Do not be too hasty. As mixing progresses permit material to ball up, roll, and break up into smaller damp particles, then to very stiff paste.) Stir additional water into the paste until the desired consistency is obtained. A 3 cu. ft. capacity concrete mixer is ideally suited for this purpose. The amount of water needed will vary according to the fineness of the dry materials. The paint should have the consistency of rich cream, except that the first coat for open-textured concrete surfaces should have a slightly thicker consistency.

2. Workability is improved by allowing the mixture to stand 20 to 30 minutes,

stirring occasionally. Most paints remain in a usable condition for 3 to 4 hours after being prepared. Some paints, especially those containing calcium chloride, should be used within 3 hours if mixed during hot weather.

3. The paint tends to stiffen during use and it is common practice to add more water when necessary. Proper consistency sometimes can be restored by vigorous stirring. To prevent solid content from settling keep paint in scaffold bucket well stirred.

(e) *Application.*

1. The two types of portland-cement paint differ in portland-cement content. Where maximum water repellency is desired, use type II (80%) portland-cement paint. Use class A paint (without filler) for all surfaces except those with very open textures. Use Class B paint (with filler) on walls with rough, porous surfaces and on such materials as cinder blocks, lightweight aggregate blocks, and similar surfaces where a heavy coating is necessary to fill voids and crevices. For some surfaces class B is used for the first coat and class A for succeeding coats.

2. Two coats are generally needed, although one is sometimes enough. The number of coats needed depends on the surface to be painted, type of paint, its coverage and waterproofing efficiency, and method of application.

3. Portland-cement paints cannot be satisfactorily applied to rough surfaces with ordinary hairbristle paint brushes. To insure proper application to coarse-textured surfaces, use a brush with relatively short, stiff, fiber bristles, such as ordinary scrubbing brushes, and autofender cleaning brushes.

4. When painting rough, porous surfaces for the dual purposes of waterproofing and decoration, scrub on paint vigorously to work it back into voids and provide a continuous film free from openings through which water might penetrate.

5. Sprayed-on paint gives less waterproofing protection than scrubbed-on coatings, therefore, spray application is not recommended.

6. Avoid excessively thick paint coatings. On the other hand, too much water in the paint, or brushing it out too thin, is equally incorrect. Such coatings look well at first but generally lose their protective and covering value much sooner than thicker coats. Examine the paint film from reading distance. If surface contains pinholes after paint is dry the paint was overthinned.

7. Do not apply paint unless temperatures are 45° F. and rising. Do not paint in exposed sunlight or during windy weather. Paint on the shady side of buildings; or if possible during cloudy, damp days or after sunset. The first 6 to 12 hours after paint has been applied is the critical period. If initial set of cement can be accomplished during this period without drying out cement film, further serious damage is not likely.

(f) *Curing.* Proper hardening of paint depends on presence of enough moisture for chemical reaction with the portland cement. After coating has reached initial set, spray the surface with a fine mist and keep in a wet condition until the second coat is applied, then continue spraying the finish coat for at least 2 additional days. Start water-spraying as soon as paint has hardened enough to resist damage by the spray, usually about 6 to 12 hours after application. Dampcuring always improves hardness and durability of paint and often means the difference between a satisfactory job and a poor one. Under certain conditions of ideal curing weather spraying of final coat may be drastically reduced or eliminated.

*Caution:* **Do not apply portland-cement paint on wood or metal; enameled, glazed, or vitrified clay products; gypsum plaster; or concrete floors.**

## 62. Adaptation From Other Military Department Publications*

*a. Introduction.* Considerable funds are being wasted annually in the purchase and misapplication of "waterproofing" materials on masonry structures. There are two basic reasons for this waste. The first is the failure of personnel to properly investigate and analyze their leakage problems, and the second is a lack of knowledge of limitations of "waterproofing" materials that are flooding the market. This section is intended to provide data on these two items. (See Maintenance Supplement No. 2 BUDOCKS Technical Digest No. 49, Pages 27–32, October, 1954, U.S. Navy Bureau of Yards and Docks.)

*b. Analyze Your Problem First.* Water entry into masonry structures can be caused by numerous conditions, many of which are difficult to determine. Often the cure is basic, but investigation of the cause, and some work, are required for correction.

*c. Passage of Water.* The paragraphs that follow illustrate in order of priority the most frequent causes of the passage of moisture through masonry walls, i.e., the first, Voids in Mortar Joints, is the most frequent cause; and second, Porosity of Masonry (which incidentally is the only cause that can be corrected solely by the application of "waterproofing" material) is the least frequent.

*d. Voids in Mortar Joints.* Voids in mortar joints are the result of soft disintegrated mortar, or the methods and quality of workmanship employed in laying up the masonry; also the extent of the bond between the mortar and the masonry units. The most common source of leakage is through voids in the vertical joints. In brick masonry, these voids usually result from the practice of simply buttering the corners of the brick, when laying, and then throwing mortar into the empty joints and only partially filling them. Correct practice is to apply a full coat of mortar to the entire end or entire side of each brick and then shove the mortar-covered end or side of the brick tightly against the brick laid previously. Under these conditions, the application of exterior coatings on the masonry will not permanently bridge such voids and early recurrence of the leaks may be anticipated. The only effective means of stopping the leaks through such voids in mortar joints is to cut the mortar joints out and repoint them with portland-cement mortar. Old mortar should be cut out by chipping or with electric driven grinding wheels.

*e. Repointing.* All loose mortar in masonry joints should be removed to solid material or to a minimum of ¾ inch depth. All such joints and all other joints in which the mortar is missing should be repointed with an approved nonshrinking mortar compound consisting of portland cement, sand, and a dry powder composed of specially prepared metallic aggregates, combined with cement-dispersing agents and reagents that will accelerate oxidation, cause a swelling of the mortar mass, and promote increased strength. Mortar mix should conform to the recommendations of the manufacturer of the metallic aggregates. Mix only enough mortar for immediate requirements because mortar with mineral aggregates should not be retempered. Thoroughly brush and blow out then wash out with water, the cutout joints. Apply mortar while the bricks are still damp. Do not use a staining metallic aggregate where discoloration is objectionable, but select a mortar of stiff consistency (par. 46*f*). For all conditions except where severe water problems are involved these mortars will give good results. Wet cure patch work as long as possible, at least two or three days. Shield from the sun in warm weather. Requirements for the metallic material are—

Waterproofing compound (metallic type) shall consist of commercial pulverized cast iron mixed with a chemical oxidizing agent. The commercial pulverized cast iron shall consist of a minimum of 85 percent by weight of metallic iron on the magnetic portion. The chemical oxidizing agent shall be sal ammoniac, calcium chloride, or sodium chloride and shall not be less than 3 percent nor more than 5 percent by weight. The iron shall contain not more than 0.05 percent by weight of oil on the magnetic portion. Iron oxide content shall not exceed 5 percent by weight. Presence of dirt, paraffin, bitumen, or other foreign substance in excess of 1.0 percent by weight of the waterproofing compound will be cause for rejection. The iron particles shall be graded within the following limits:

| Screen mesh | Percentage retained |
| --- | --- |
| 35 | 0– 5 |
| 40 | 0–10 |
| 60 | 0–35 |
| 100 | 30–55 |
| 200 | 75–90 |

---

* This portion of section XI is adapted from other military department publications. It shows slight variations and repetitions of information previously given in this manual; however, because of the controversial nature and importance of Waterproofing and Dampproofing it is included here to indicate their solutions to these problems.

*f. Small Cracks.* Where only shrinkage cracks are present (cracks less than ⅛ inch in width) but the mortar is otherwise sound, a dry wall can be produced by the application of exterior cement water paint conforming to Federal Specification TT-P-21. Saturate with water the masonry surface and fill with the cement water paint by working it into the cracks with a stiff brush. Mix cement water paint to a heavy creamy consistency.

*g. Large Cracks.* Cracks in masonry walls may be the result of one or more of the following defects; separately or in different combinations.

(1) Unequal settlement or distortion of the footings or structure supporting the masonry.

(2) Overstressing of the masonry caused by external forces of unexpected severity which may occasionally develop from earth pressure, earthquakes, hurricanes, and the like.

(3) Abnormal distortions, especially expansion, of appurtenant elements such as floors and roofs.

(4) Volume changes in the masonry involving principally temperature shrinkage and moisture shrinkage.

Except for small surfaces cracks which do not extend through the wall, leaks resulting from cracks caused by the above, impose problems other than leakage. To avoid recurrence of the leak, correct the basic causes first, then cut back crack to form a V-shaped groove to a depth about equal to the width of the **V** at the surface, but not less than ½ inch, and then sealed by pointing, following the instructions contained in paragraphs 60*d* or 61*a*(4). If cracks are due to settlement of the foundation they should be repointed with masonry mortar. These usually occur near corners of buildings and form a pattern of "step-like" horizontal and vertical cracks. Cracks due to expansion either from moisture, temperature, or delayed hydration of magnesium oxide (MgO) in the mortar result in the longitudinal movement of the wall toward the ends, and slippage at or near the foundation. They are evidenced by the presence of wedge-shaped cracks above and below windows as well as large adjacent vertical cracks. The major pattern of these cracks should be filled with calking compound and minor patterns with cement mortar. In extreme cases where thermal action is the principal cause of expansion cracks, additional vertical expansion joints should be sawed into building walls. (See technical paper "Thermal Expansion in Clay masonry Structures," March,

1958, Volume g, No. 3 of Structural Clay Products Institute, Wash. 6, D.C.—copy upon request.) When cracks are less than ⅛ in. wide, seal by the same method used for shrinkage cracks. Wet cure the patch work as long as possible, at least two or three days, and in warm weather protect it from rays of the sun. Fill cracks around window sashes or door frames with oakum to within ½ inch or ¾ inch of the outside surface and then finish with a calking compound conforming to Fed. Spec. TT-C-598 or TT-S-00227.

*h. Importance of Repairs.* Unless such repairs are completed first, the application of coatings to the exterior of the wall will provide, only the most temporary relief. If repairs are properly completed, they should stop the leaks and exterior coatings will be unnecessary.

*i. Flashing Failures.* Flashing failures are frequently responsible for the presence of moisture in a well-constructed masonry wall. This is especially true when the wall includes a roof parapet, where cause of moisture entry can be traced to failures at the juncture of the roof and parapet walls. It is apparent that in these instances, the solution lies in repair of the flashing rather than in application of exterior coatings on the masonry.

*j. Condensation.* Condensation is frequently mistaken for a leak in a masonry wall. Exterior walls of heated buildings, located in cold climates or in areas subjected to cold weather during a portion of the year, may appear to be leaking as a result of water passing through the walls, even though the walls are watertight. In such cases, investigation will usually disclose that when relative humidities within the building are comparatively high, exterior relative humidity rather low, and exterior temperatures are below the freezing point, moisture vapor tends to travel through the wall, from the interior to the exterior. The temperature of the masonry wall is considerably lower than that of the heated interior and the temperature of the exterior portion of the wall tends to approach that of the exterior atmosphere. Moisture vapor traveling from the interior to the exterior encounters the low temperatures within the wall, and when the dew point temperature of the vapor is reached, the moisture within the vapor condenses and a damp wall results. Under certain conditions, freezing and thawing may occur within the walls, causing cracking and spalling of the masonry and actual leakage.

*k. Vapor Barrier.* The above conditions may be prevented by providing a continuous vapor barrier on the interior face of the wall, or by providing sufficient ventilation within the building to prevent the development of high humidities. The most satisfactory vapor barrier is aluminum foil but this is frequently impractical to install. The following paints act, to some extent, as a vapor barrier and may be used: semi-gloss white and light tints conforming to Federal Specification TT–E–508; gloss white and light tints conforming to Federal Specification TT–E–506; gloss-black and dark colors, Federal Specification TT–E–489; and chlorinated rubber paint, white and light tints, Federal Specification TT–P–95; and latex base paint, Federal Specifications TT–P–29.

*l. Porosity of the Masonry.* Porosity of the masonry is the only cause of leaks that can be corrected solely by the application of a coating to the wall. If it is determined beyond a reasonable doubt that porosity is the only cause of the leak, the application of a suitable coating on the wall will correct the trouble. Apply the coating to the exterior walls; if applied to the interior face, it usually results in blistering. Under certain conditions, such as basement walls (app. IV), manhole walls, and other structures where it is impractical to expose the exterior of the wall, special treatment of the interior wall will provide satisfactory results; but this is only recommended when the exterior wall is inaccessible.

*m. Selection of Coatings.* In the repair of leakage due to porosity of the masonry, care must be taken in the selection of the proper coating. The market is flooded with cure-all "waterproofing" compounds, most of which are highly priced. Many of the older line proprietary coatings are nothing but Federal Specification TT–P–21 at a fancy cost. Many of the newer transparent ones are "secret" formulas, most of which fail to hold up. Of the more recently developed coatings, the silicone water repellents have the most promise. However, make certain that the silicone content is adequate. Many products are formulated with only 1 percent or 2 percent of silicones, which is inadequate.

*n. Coatings for Porous Masonry—Exterior Application.* The following coatings will give satisfactory results when applied to exterior walls above grade, provided porosity of the masonry is the only cause of water entry:

(1) Cement-water paint conforming to Federal Specification TT–P–21. Properly applied this material will give excellent service. It may be applied over previously unpainted concrete and masonry surfaces or even surfaces covered with the same type of paint. In order to be effective in resisting the entrance of water, apply the paint with stiff brushes.

(2) Oil-paint for concrete and masonry conforming to Federal Specification TT–P–24. This material may be applied to previously unpainted concrete and masonry surfaces, without the use of a primer; or to surfaces coated with cement-vapor paint.

(3) Silicone water repellents having a minimum nonvolatile content of 5 percent silicone resin dispersed in zylene, or toluene provide a "colorless" transparent finish. This material permits breathing. Percentage of silicone resin and the formulation of the compound vary with each manufacturer. Repellent should comply with Federal Specification SS–W–00110.

*o. Coatings for Masonry Surface—Interior Application.* Where exterior walls are inaccessible and coating must be applied to the interior face, the following is recommended:

Expose, clean and roughen wall to provide a key for the waterproofing materials. Cut back holes, cracks, and other soft or porous places to solid material; clean; point with mortar, and calk all pipes, bolts, and similar construction with lead wool and waterproof cement; and make watertight. Drench surfaces with clean water and give one bonding coat composed of 1 part of cement and 1 part of metallic material mixed to a creamy consistency, and apply with fiber brushes. Thoroughly brush to seal all pores, rather than merely provide a surface veneer. This is followed by two coats of mortar, composed of 1 part of portland cement, 3 parts of sand, and 25 pounds of metallic material to each bag of cement. Trowel the first coat on and scratch when partially dry; and brush on the second coat (mixed to a heavy brushing consistency) carefully and float with wood floats. Total thickness of all coats should be approximately $\frac{5}{16}$ inch to $\frac{3}{8}$ inch. After each coat has set but not dried out, wet surfaces frequently over a period of at least 72 hours. Allow sufficient time between coats to permit thorough oxidation of the material. Instructions on use of metallic material are given in paragraphs 61a(4) and 62e. Repeating, mix only enough mortar for immediate use because it should not be re-tempered.

*p. General Data on Waterproofing Compounds.* To assist personnel in evaluating the claims of the various manufacturers of waterproofing materials, and to assist them when procurement of proprietary material is necessary, the following supplemental general information is provided. Almost anything can be called "waterproofing" if it is qualified in fine print. The terms "waterproofing" and "dampproofing" have been used indiscriminately in the industry. This has resulted in misunderstandings. In an effort to clarify the situation, the Federal Trade Commission issued a set of "Trade Practice Rules" for the Masonry Waterproofing Industry under date of 31 August 1946. These rules regulate, among other things, the use of such terms as waterproof, dampproof, water-resistant, damp-resistant, weatherproof, watertight, moistureproof, vaporproof, and the like.

(1) Under Section (III) of Rule 1, it is stated that "under this rule, no representation shall be made directly or by implication, unless such be wholly true and nondeceptive, that any product will be fully effective regardless of—

(a) The kind, porosity or condition of masonry units or masonry structures with which it is integrated, or to which it is applied;

(b) The manner of its integration or application;

(c) Cracks or other structural failure resulting from ground movement, settlement, or other cause;

(d) The geographical location of the structure to which it is applied, or the water, moisture, or atmospheric conditions which such structure may encounter."

(2) Under Rule 2, it is stated that—

"(a) In the sale, offering for sale, or distribution of industry products, it is an unfair trade practice to use the word "waterproof," "waterproofing," or any other word or representation of similar import, as descriptive of any industry product unless, when properly integrated with or applied to masonry units or masonry structures, the product will render such units and structures, impermeable to, or proof against, the passage of water and moisture throughout the life of such units, or structures, and under all conditions of water or moist contact or exposure; subject, however to the following further provisions of this rule:

(b) If an industry product will effect impermeability for a substantial period of time although less than, or not known to be as long as the life of the masonry or structure; or, if for a substantial period of time the industry product will render some but not all types of masonry units or structures impermeable to water and moisture; or if for a substantial period of time the industry product will render masonry units or structures impermeable to water under some but not all conditions applicable in the use of industry product or the function of such masonry units or structures, then the words "waterproof," "waterproofing," or representations of like import, may be used as descriptive of said industry product provided such words, terms, and representations are accompanied and qualified, in immediate conjunction and in accordance with the following provisions, by a clear, conspicuous, and nondeceptive statement of limitations or disclosure showing the respective applicable limitations: of the period of time impermeability will endure; of the type or kind of masonry unit or structure in which the product will produce such impermeability; and of the other applicable conditions which need be present or absent in order for said result of impermeability to be achieved and to endure during said specified period of time."

In other words, read the manufacturer's literature carefully, particularly the fine print showing the respective applicable limitations.

# SECTION XII
# CLEANING

## 63. Methods

*a. General.* Materials to be cleaned and the type of spots or stains to be removed determine the cleaning process used. Sandblasting and acid-washing work well on some brick, concrete, granite, and other hard surfaces, but are harmful to soft stone surfaces and glazed finishes. A method that cleans marble or glazed surfaces successfully may be entirely unsatisfactory on concrete or rough-textured clay-brick surfaces. Safe methods for a masonry surface may damage mortar joints between masonry units. If the surface has more than one kind of masonry, one cleaning method may not be safe or effective on all.

*b. Concrete.* Many chemicals can be applied to concrete and concrete masonry without much injury, but chemicals having an acid reaction should not be used. Even weak acids may roughen the surface if left on it long. Stains on concrete and masonry caused by iron, copper, bronze, aluminum, ink, tobacco, urine, fire, lubricating oil, rotted wood, coffee, and iodine, and general service stains can usually be removed by the methods listed below. Old, long-neglected stains may require repeated treatments. Remove deep stains with a poultice or bandage treatment. Make the poultice by mixing active chemicals with fine inert powders to a pasty consistency and apply a thick layer to the stained surface. Prepare the bandage by soaking cotton batting or several layers of cloth in chemicals; paste bandage over the stain. More detailed methods of stain removal are described later in this chapter.

*c. Masonry.* Masonry units of low absorption and those with smooth or glazed surfaces generally respond readily to proper cleaning methods and resume their original appearance. Highly absorbent or rough-surfaced units are more difficult to clean. If the stain has penetrated the pores of the unit, cleaning may remove part of the surface, destroy its texture, and change its appearance. Principal methods of cleaning masonry structures are with steam, water, sandblasting, and various liquids and pastes.

*d. Scaffolds.* Use either a painter's overhead, suspended type or stationary ground-supported scaffold when cleaning surfaces high above ground. The kind used depends on building height and other local conditions (par. 68).

## 64. Steam Cleaning

Cleaning with high-pressure steam and water, sometimes called cold-steam cleaning, is effective and economical. It removes grime from weather-exposed surfaces of concrete and masonry buildings without harming surfaces or preventing later weathering. After steam cleaning, masonry surfaces retain their original finish and natural color tones without the roughened surfaces and dulled arrises caused by sandblasting or the dead, bleached tone caused by acid cleaning. Steam cleaning equipment including boiler must be inspected and certified by qualified personnel prior to start of operations. Provisions of the ASME Boiler Code must be complied with. Proposed cleaning and operating procedure must be carefully evaluated to assure continued safe operating conditions.

*a. Equipment.* Use proper equipment for steam-cleaning. Much of it can be improvised, but needs of the post may make buying or renting equipment advantageous.

(1) *Steam supply.* See that there is a continuous supply of high-pressure steam or water vapor before cleaning operations begin. A portable boiler with accessories, truck-mounted for easy transportation, is generally a satisfactory supply source. Steam pressure for cleaning old buildings is preferably 150 pounds and *never* less than 140 pounds. For cleaning new work, 120 pounds, *never* less than 100 pounds, are needed. Boiler capacity of about 12 horsepower for each cleaning nozzle is necessary. The pressure with which the steam and water mixture is driven against the wall, not volume of discharge, is important in steam cleaning.

(2) *Nozzle.* The cleaning nozzle is a most important accessory. It should be a mixing type having a water-control valve and automatic steam shut-off. Several nozzles are available. One efficient type has a very narrow opening 4 inches long and can deliver an extremely fine spray at a high velocity. Operating two cleaning nozzles from each length of scaffold is good practice.

(3) *Water-supply hose.* Ordinary garden hose is suitable for carrying water from source of supply to scaffold. Use only high-pressure steam couplings, with suitable pipe and fittings, if pipe is used to convey steam from boiler to scaffold.

(4) *Rinsing hose.* In addition to the hose supplying water to the steam-cleaning nozzle, have another hose (ordinary garden-type nozzle with shut-off) to flush the walls with water occasionally.

b. *Procedure.*

(1) *Cleaning.* Cleaning is done by high-velocity projection of a finely divided spray of steam and water against the masonry surface. The mixture of steam and water spray entering minute surface depressions and openings dissolves and dislodges grime, soot, and other extraneous matter, which is later flushed down the wall by the rinsing hose. Experimentation with equipment used quickly shows the operator the best angle and distance of nozzle from the wall and proper regulation of steam and water valves for most effective work. Work on one 3-foot-square space at a time. Pass the nozzle back and forth over the area, then flush with clear water before moving to the next space.

(2) *Cleaning additives.* Alkalies such as sodium carbonate, sodium bicarbonate, and trisodium phosphate are sometimes added to the cleaning water to speed the cleaning action. While these salts aid in cleaning, some are retained in the masonry, and may appear on the wall later as efflorescence. To cut down the amount of salts retained, wet wall units thoroughly with clear water before beginning the cleaning operations. Immediately after cleaning, wash the wall with plenty of clear water to remove all possible salts from the face of the wall.

c. *Removing Stains.* Removal of surface dirt sometimes reveals stains. A mild acid wash may be necessary to remove them. After the stain is removed, steam the treated surface again and flush with water from the rinsing hose to remove all trace of acid wash (par. 58).

d. *Removing Hardened Deposits.* Steel scrapers or wire brushes may be necessary to remove hardened deposits that cannot be removed by steam cleaning.

Use scrapers sparingly and carefully to avoid damage to masonry surface. Use brushes of fine spring-steel wire to grind off the hard deposit without digging into or scratching the cleaned surface. After the hard deposit has been removed by scraping or wire brushing, flush off the surface with water, treat it with the steam-cleaning nozzle, and flush again to remove all loose dirt which might cause future streaking or discoloration.

## 65. Sandblast Cleaning

Sandblasting cleans rust and scale from structural-steel members, bridges, gas holders, oil tanks, and many other metal surfaces efficiently. However, do not use this method on marble, terra cotta, glass or units with glazed or other special surfaces or textures. Although it cleans effectively, it often destroys the original surface of the masonry unit. It tends to dull sharp edges, blur ornamental detail and carving, and roughen surfaces. Stone cleaned by sandblasting, especially limestone, appears whiter, but this whiteness is caused partly because the stone surface has been bruised. In sandblasting, compressed air forces hard sand through a nozzle against the masonry surface. The sand removes accumulated grime and a thin layer of masonry. For best results follow the procedures listed below.

a. *Hose.* Use a ¾-inch hose if maximum pressure of air and sand is required to remove dirt, and a 1-inch hose if volume rather than high pressure is desired.

b. *Sand.* Quality of sand needed varies with depth of cutting done. Placing a canvas screen around the scaffold platform keeps sand from scattering and makes it possible to salvage about 75 percent of the sand.

c. *Personnel.* Usually, four men operate a sandblasting outfit; one at the machine, one at the nozzle, and two on the ground to handle hose, scaffold, sand, and so on.

d. *Repointing Mortar Joints.* When hardness between masonry units and mortar joints differs widely, sandblasting may cut too deeply into the mortar. If this occurs, joints must be repointed.

e. *Waterproofing.* Because roughened surfaces produced by sandblast cleaning quickly gather soot and dirt, an application of transparent waterproofing is desirable. This coating fills surface pores, tends to make the wall self-cleaning, and prevents rapid soiling of the surfaces by smoke and dust (par. 61b(1)).

## 66. Acids, Caustic Washes and Paste Cleaners

a. *Selection of Cleaner.* Acid, caustic concentrations, and paste mixtures are used to clean many interior and exterior concrete masonry surfaces. The cleaning material used depends on the type of surface to be cleaned, age of structure, and material or stain to be removed. The most satisfactory results can be obtained by developing a cleaning material for the job at hand after analyzing the dirt and stains to be removed. Strength and chemical composition of the cleaner can usually be adjusted by observing the action on trial spaces. Protect grass and shrubbery from damage.

b. *Cautions in Using Acids.*

(1) Do not use acid solutions to clean limestone and similar materials unless experienced operators and expert supervision are available. Acid washes tend to eat into the stone surfaces and pit them. They usually bleach, producing an unnatural appearance, and may cause yellow stains to appear later. If an acid solution is not thoroughly washed from the masonry pores after cleaning, the destructive action continues for some time.

(2) When mixing acid solutions, always pour acid into water.

(3) Handle acid solutions carefully because they are harmful to the skin and especially to the eyes. When working with acid cleaners, wear goggles, gloves, and protective clothing, and keep a supply of running water at hand.

(4) Use only wooden containers and fiber brushes when cleaning with acids. Do not use wire brushes or steel wool to scrub the walls, because small steel particles become lodged in crevices, producing rust spots and stains.

c. *Soap Powder.* Many burned-clay and concrete surfaces and glazed and polished surfaces of tile, marble, and glass can be cleaned by hand-scrubbing with a white soap powder dissolved in soft water, using ordinary fiber scrub brushes. Wash surfaces thoroughly with clear water after scrubbing them.

d. *Mortar Stains.* To clean mortar stains, first remove excess mortar with a putty knife or a chisel. Soak the wall with clear water and wash it by applying one of the following acid solutions with a fiber brush or broom:

Ten parts water to 1 part hydrochloric (muri-

atic) acid Federal Specification O–H–765, technical grade, 31 percent solution.

Twenty parts water to one part phosphoric acid, commercial grade, 75 percent solution.

Since acid solutions attack mortar joints, wash the wall thoroughly with clear water immediately after it is cleaned. Remove final traces of acid by applying a dilute ammonia solution (1 pint of ammonia to 2 gallons of water).

e. *Efflorescence.* Water applied with stiff scrubbing brushes frequently removes efflorescence. If this does not remove all stains, follow the procedures described in d above, using a water-hydrochloric acid solution.

f. *Stains on Cement Stucco.* To clean white cement stucco, wet the surface, apply a solution of 20 parts water and 1 part sulfuric acid (commercial grade, 93 percent solution) and rinse thoroughly with clear water. Hydrochloric acid solutions may produce a yellowish tinge on white cement.

g. *Paints and Similar Coatings.*

(1) *Whitewash, calcimine, and cold-water paints.* To remove whitewash, calcimine, and cold-water paints, wash the surface with an acid solution of one part muriatic acid and five parts water. Scrub vigorously with a fiber brush as the solution forms. When the coating has been removed, wash the wall with water from an open hose until all trace of acid is removed.

(2) *Oil paints, enamels, varnishes, shellacs, and glue sizings.*

(a) Remove oil paint, enamel, varnish, shellac, or glue sizing by applying a paint remover, leaving it on until softened paint can be scraped off with a putty knife, or flushed off with water. After using paint remover, wash the wall thoroughly to remove all traces of acid. Efficient paint removers include—

1. An acceptable noncombustible, low toxicity paint remover.

2. Two pounds trisodium phosphate in 1 gallon hot water.

3. Two and 1/2 pounds caustic soda in 1 gallon hot water.

4. One part sodium hydroxide dissolved in three parts water and added to one part mineral oil. Stir mixture until emulsified, then stir in one part sawdust or other inert material.

5. Equal parts soda ash and quicklime mixed with enough water to form a

thick paste. Leave this mixture on the wall for 24 hours after application, then scrape off.

(b) If the oil-paint film is very thick and hard, sandblasting may be the best way to remove it. Burning is sometimes used, but is not recommended because of the fire hazard.

(c) If a paint film is old, crumbling, and flaking, scraping it off with wire brushes and metal scrapers may be necessary. While this method is effective, it may leave metal particles which later cause rust stains in the wall surface.

## 67. Cleaning of Miscellaneous Stains

Use one or more of the following materials and methods to remove stains from clay, concrete masonry, stone and marble:

*a. Iron Stains.* Iron stains can usually be recognized by their resemblance to iron rust or by their proximity to steel or iron members in the building. Large areas of concrete or cement stucco may be stained if curing water used contains iron. Remove by mopping the surface with a solution of 1 pound of oxalic acid dissolved in 1 gallon of water. After 2 or 3 hours, rinse with clean water, scrubbing at the same time with stiff brushes or brooms. Some spots may require a second mopping and scrubbing. For older, deeper stains the following methods are recommended:

(1) *Method 1.* Dissolve one part sodium citrate in six parts lukewarm water. Mix thoroughly with seven parts of lime-free glycerine. Add to this solution enough whiting or kieselguhr to make a paste poultice stiff enough to adhere to the surface when applied with a putty knife or trowel to a thickness of ¼ inch or more. Allow a minimum of 2 days for drying. Scrape off and wash thoroughly. If the stain has not disappeared, repeat the treatment. This treatment has no injurious effects, but its action may be too slow for bad stains. Ammonium citrate produces quicker results, but may injure a polished surface slightly, making a repolish job necessary.

(2) *Method 2.* The sodium hydrosulphite combination bandage-poultice method is more satisfactory for removing deep, intense, iron stains.

(a) Make a solution by dissolving one part sodium citrate crystals in six parts of water. Dip white cloth or cotton batting in this solution, place the cloth over the stain, and leave it there for 15 minutes.

(b) On horizontal surfaces, sprinkle a thin layer of hydrosulphite crystals over the stain being treated with sodium citrate, moisten with water, and cover with a paste of whiting and water.

(c) Give vertical surfaces the sodium-citrate treatment. Place a layer of whiting paste on a plaster's trowel, sprinkle on a layer of hydrosulphite crystals, moisten slightly, and apply it to the stain. Remove treatment after 1 hour. If it is left on longer, a black stain may develop. Wash treated surface with clean water. If inspection shows incomplete removal of the iron stain, repeat the cleaning operation, using fresh materials.

*b. Tobacco Stains.*

(1) *Method 1.* Dissolve 2 pounds of trisodium-phosphate crystals in 4 to 5 quarts of hot water. Mix 12 ounces of chlorinated lime to a smooth stiff paste in a shallow enameled pan by adding water slowly and mashing the lumps. Pour this and the trisodium-phosphate solution into a 2-gallon stoneware jar and fill it with water. Stir well, cover the jar, and allow lime to settle. Add some of the liquid to powdered talc, stirring until a thick paste is obtained. Apply the paste with a trowel to a poultice thickness of ¼ inch. Scrape off the dry paste with a wooden paddle or trowel. This mixture is a strong bleaching and corrosive agent. Care must be taken not to drop it on colored fabrics or metal fixtures.

(2) *Method 2.* Minor tobacco stains can be removed by applying ½-inch-thick poultices of a stiff paste made of water and any of the several grit scrubbing powders commonly used on marble, terazzo, and tile floors. Scrape off the dry paste with a wooden paddle and wash the surface with clean water. In most cases, two or more applications will be necessary.

*c. Fire and Smoke Stains.*

(1) *Method 1.* Fire and smoke stains can sometimes be removed by scouring with powdered pumice or a grit scrubbing

powder. After removing the surface stain by scouring, the deep-seated stains can be removed by applying the trisodium-phosphate-chlorinated-lime solution described above for tobacco stains. Fold a white canton flannel cloth to three or four thicknesses and saturate it with the liquid. Paste this saturated cloth over the stain and cover it with a slab of concrete or sheet of glass, making sure the cloth is pressed firmly against the stained surface. If the surface is vertical, devise a method to hold the saturated cloth firmly against the stain. Resaturate the cloth from time to time. Wash the surface thoroughly at the end of the treatment.

(2) *Method 2.* Make a smooth stiff paste of trichlorethylene and powdered talc, and apply it as a troweled-on poultice. Cover the poultice with glass or a pan to prevent rapid evaporation. Allow time to dry. Scrape off and wash away all traces of treatment material. Trichlorethylene gives off harmful fumes, therefore, see that closed spaces are well ventilated when using this stain remover.

*d. Copper and Bronze Stains.* These stains are nearly always green, but in some cases may be brown. Mix one part dry amonium chloride (sal ammoniac) and four parts powdered talc, add water, and stir to a thick paste. Trowel ¼-inch layer of paste over the stain and leave until dry. When working on polished marble or similar fine surfaces, use a wooden paddle to scrape off dried paste. An old stain may require several applications. Aluminum chloride may be used instead of sal ammoniac.

*e. Aluminum Stains.* These appear as a white deposit which can be removed by scrubbing with a 10 to 20 percent muriatic acid solution. On colored concrete, use the weaker solution. Wash thoroughly with clean water.

*f. Oil Stains.* Oils penetrate most concrete readily. Oil spilled on horizontal surfaces should be covered immediately with a dry powdered material such as hydrated lime, Fuller's earth, or whiting. Sweep up the powdered material, taking as much of the oil as possible. Scrub with a solution of 1.1.1 trichlorethane, technical, inhibited (methyl chloroform) Federal Specifications O–T–620. If the treatment is made soon enough, there will be no stain. However, when oil has remained for some time, one of the following methods may be necessary:

(1) *Method 1.* Mix 1 pound of trisodium phosphate in 1 gallon of water and add sufficient whiting to make a stiff paste. Spread a layer ½ inch thick over the surface to be cleaned. Leave paste until it dries (about 24 hours), remove, and wash with clear water.

(2) *Method 2.* Saturate white canton flannel in 1.1.1 trichloroethane, technical, inhibited (methyl chloroform) Federal Specification O–T–620, and place it over the stain. Cover the cloth with a slab of dry concrete or sheet of glass. If stain is on a vertical surface, improvise means to hold cloth and covering in place. Keep the cloth saturated until the stain is removed. Covering saturated cloth with glass tends to drive the stain in, while the slab of dry concrete will draw out some of the oil.

*g. Coffee Stains.*

(1) Saturate a cloth in a solution of one part of glycerin and four parts of water. Place the cloth over the stain and resaturate the cloth from time to time until the stain disappears.

(2) Javelle water (sodium hypochlorite) can be used as the bleaching agent instead of the glycerin and water solution. Javelle water can usually be purchased at drug stores. If not, it is prepared as follows: Dissolve 3 pounds of common washing soda in 1 gallon of water. Mix 12 ounces of chlorinated lime to a paste in a shallow enameled iron pan by adding water slowly and mashing the lumps. Add the paste to the soda solution, pour the mixture into a stoneware jar, and add water to 2 gallons. Stir thoroughly, cover jar, and allow lime to settle. Draw off the clear liquid and dilute it with six times its volume of clear water. Use as a soap or scrubbing solution, rinsing the surface thoroughly before and after application. Javelle water is a strong bleaching material and should not be allowed to drop on colored fabrics. It is not recommended for general cleaning purposes.

*h. Iodine Stains.* Iodine stains gradually disappear of their own accord. Hasten removal of stains from horizontal surfaces by applying alcohol covered with whiting or talcum powder. If stain is on a vertical surface, make a paste of talcum and alcohol, brush alcohol on the stain, and cover with the paste.

Allow paste to dry, scrape it off and wash surface with clear water.

*Caution:* **Alcohol is highly flammable, therefore only a small area should be treated at a time, and fire extinguisher should be handy for use, if required.**

*i. Perspiration Stains.* Secretions from the hands and oils from the hair produce brown or yellow stains that may be mistaken for iron stains. Removal methods described for fire and smoke stains are recommended. Bad stains may require several treatments.

*j. Urine Stain.* Use the method described for tobacco stains. Should the stain prove stubborn, saturate cotton batting in one of the liquids, paste it over the stain, and resaturate it from time to time until the stain is removed. Complete the treatment by washing with clear water.

*k. Ink Stains.* Stains made by different inks may require dissimilar removal treatments.

(1) The acid content of many ordinary inks generally causes an etching action on concrete, masonry, and marble surfaces. Prompt removal should be attempted. Make a strong solution of sodium perborate in hot water. Mix with enough whiting to make a thick paste, apply a $\frac{1}{4}$-inch layer over the stain, and leave until dry. If some blue color remains after poultice is removed, repeat the treatment. If a brown stain remains, treat it by method *l* for iron stains.

(2) Many bright-colored inks are water solutions of synthetic dyes. The sodium-perborate poultice generally removes these stains. Other removal treatments are—

(a) Cover stain with a cotton-batting bandage saturated with ammonia water.

(b) Use Javelle water in the same way as ammonia water, or mix it with whiting to make a thick paste and apply a layer over the stain.

(c) Use as a poultice a mixture of chlorinated lime and whiting reduced to paste and water.

(3) Some blue inks containing prussian blue, or ferrocyanide of iron, produce a stain which cannot be removed by the perborate or chlorinated lime poultices, or the Javelle-water treatments. Such stains can be treated by covering them with cotton batting saturated with ammonia water. Strong soap solutions applied in the same way may also produce satisfactory results.

(4) Indelible inks often consist entirely of synthetic dyes. These stains may be treated as described in (1) above. Some indelible inks may contain silver salts which produce black stains. These stains can be removed by applying bandages saturated with ammonia water. Usually several bandage applications are necessary.

# SECTION XIII

# SAFETY

## 68. Safety Measures

Section III of TM 5–610 is an outline reference on matters pertaining to the prevention of accidents. Full compliance with above manual is required. Some of the main safety provisions to remember are as follows:

*a.* Comply with safety provisions for excavation, trenching, trench and bank shoring.

*b.* Follow safety provisions when building and using ladders, runways, ramps, platforms, stationary and swinging scaffolds, temporary stairs, guardrails, and other facilities (par. 63*d*).

*c.* See that all machinery and other equipment is inspected by a qualified person and approved as in safe operating condition before it is put in use. Fiber and metal ropes, especially those used for supporting workmen, must be inspected immediately before being used and frequently during use.

*d.* Enforce safety provisions for stacking and storing new and used materials, and for disposing of waste. Keep stairways, corridors, and work spaces clear of tools, tripping hazards such as loose board, and foot hazards such as protruding nails in scrap lumber.

*e.* Do not permit workmen to ride on material hoists, hooks, and slings.

*f.* Do not permit men to work under swinging loads, booms, and buckets. Use tag lines and/or pike poles to guide a swinging load.

*g.* See that steam hoses and connectors are securely lashed in place. This prevents the hose from snaking and scalding workmen if a connection breaks. Boilers must be inspected and approved before being put in use.

*h.* Wash eyes immediately with clear water to remove lime or cement dust, caustics, and acids. Secure medical attention by a competent physician as soon as possible.

*i.* Do not allow smoking, open-flame lighting, or the use of sparking electrical apparatus when working with flammable liquids or materials that create explosive gases. Closed spaces must be well ventilated. In the absence of satisfactory ventilation, men working in closed spaces must use approved air-line respirators.

*j.* Insure that workmen wear proper protective clothing, gloves, goggles, face shields, helmets, respirators, and similar approved protective devices when working with steam, flammables, acids, corrosives, and during operations producing dust, gases, flying chips, splinters, and similar hazards.

AGO 10179A

# APPENDIX I

# REFERENCES

## 1. Federal Specifications[1]

| | |
|---|---|
| O-C-105 | Calcium Chloride. |
| O-H-765 | Hydrochloric (muriatic) acid. |
| O-S-605a | Sodium silicate, liquid. |
| O-T-620 | Trichloroethane. |
| HH-C-00466 | Fabric, glass fiber. |
| QQ-B-101c | Bases, metal, for plaster and stucco construction. |
| QQ-S-632 | Steel bar, reinforcing (for) concrete. |
| SS-A-281 | Aggregate (for) portland-cement concrete. |
| SS-B-663 | Brick; concrete. |
| SS-C-153 | Cement, bituminous, plastic. |
| SS-C-181 | Cement; masonry. |
| SS-C-192 | Cement, portland. |
| SS-C-618 | Concrete, ready mixed. |
| SS-L-351 | Lime; hydrated (for) structural purposes. |
| SS-P-371a | Pipe, concrete, non-reinforced, sewer. |
| SS-P-381 | Pipe, pressure-reinforced concrete, pretensioned reinforcement (steel cylinder type). |
| SS-P-375 | Pipe, concrete, reinforced, sewer. |
| SS-Q-351 | Quicklime, for structural purposes. |
| SS-R-531 | Roofing slabs; concrete; precast |
| SS-R-501 | Roofing; prepared; asphalt, smooth-surfaced. |
| SS-S-721 | Stone; architectural, cast. |
| SS-T-308(3) | Tile, ceramic; floor, wall and trimmers. |
| SS-T-321(1) | Tile, structural, clay, floor. |
| SS-W-00110 | Repellent, water, silicone. |
| TT-C-306 | Chromium-oxide-green; dry (paint pigment). |
| TT-C-598(2) | Compound, calking; plastic (for masonry and other structures). |
| TT-E-489b(1) | Enamel; gloss, synthetic (for exterior and interior surfaces). |
| TT-E-506c | Enamel, tints and white, gloss, interior. |
| TT-E-508 | Enamel; interior, semi-gloss, tints and white. |
| TT-I-511 | Iron oxide; red, synthetic, dry (paint pigment). |
| TT-I-698 | Iron oxide; black, synthetic, dry (paint pigment). |
| TT-I-702 | Iron oxide; brown, synthetic, dry (paint pigment). |
| TT-P-21 | Paint, cement-water, powder, white and tints. |
| TT-P-24a | Paint, oil, concrete and masonry, exterior, eggshell. |
| TT-P-29 | Paint, latex base, interior, flat white and tints. |
| TT-P-95(1) | Paint, rubber-base (for swimming pools). |
| TT-S-176 | Sealer, surface, varnish type, floor, wood or cork. |
| TT-S-00227 | Sealing compound; rubber base, two component (for calking, sealing, and glazing in building construction). |
| TT-U-450 | Ultramarine blue; dry (paint pigment). |
| TT-Y-216 | Yellow-iron-oxide, hydrated, synthetic, dry (paint pigment). |

## 2. Military Specifications[2]

MIL-D-23003 Deck covering compound, nonslip; lightweight (epoxy).

## 3. National Bureau of Standards Reports[3]

BMS-52, Water Permeability of Walls Built of Masonry Units.

Research Paper 1015, Wick Test.

BMS-95, Tests of Cement Water Paints and Other Waterproofings.

NBS Report 3079, Requirements for Concrete-Masonry. Construction (Revision of NBS Report 2462) by Cyrus C. Fishburn, 30 January 1954.

## 4. Portland Cement Association Publications[4]

| | |
|---|---|
| S8-2 | Concrete Floor Finishes. |
| CP-12 | Repairing Damp or Leaky Basements in Homes. |
| ST-1-2 | Painting Concrete. |
| ST-4-2 | Effect of Various Substances on Concrete and Protective Treatments. |
| ST-6 | Notes on Efflorescence. |

[1] Available from Superintendent of Documents, Government Printing Office, Washington, D.C. (See apps. II through V for other Federal Specifications.)

[2] Available from Superintendent of Documents, Government Printing Office, Washington, D.C.

[3] Available from Superintendent of Documents, Government Printing Office, Washington, D.C.

[4] Available from Portland Cement Association, 33 W. Grand Avenue, Chicago 10, Ill.

| ST–11 | Bonding Concrete or Plaster to Concrete. |
| ST–17–2 | Shotcrete. |
| ST–19 | Removing Stains from Concrete. |
| ST–37–3 | Surface Treatments for Concrete Floors. |
| ST–50 | Restoring Old Stone Masonry by Pressure Grouting. |
| SCB–7 | Pavement Maintenance Practices and Uses for Soil-Cement at Army and Navy Facilities. |
| HB–17–3 | Maintenance Practices for Concrete Pavements. |
| HB–18–8 | The Elimination of Pavement Sealing by Use of Air-Entraining Portland Cement. |

## 5. American Society for Testing Materials Publications[5]

ASTM Standards in Building Codes.

ASTM Standards on Mineral Aggregates and Concrete.

ASTM Standards on Cement.

## 6. American Concrete Institute Publications[6]

Building Code Requirements for Reinforced Concrete (ACI–318).

Manual of Standard Practice for Detailing Reinforced Concrete Structures—(ACI–315).

Reinforced Concrete Chimneys (ACI–505).

Winter Concreting (ACI–604).

Selecting Proportions for Concrete (ACI–613).

Measuring, Mixing, and Placing Concrete (ACI–614).

Applications of Portland Cement Paint (ACI–616).

Precast Concrete Floor and Roof Units (ACI–711).

Application of Mortar by Pneumatic Pressure (ACI–805).

Concrete Formwork (ACI–347).

Hot Weather Concreting (ACI–605).

## 7. American Standard Association Publication[7]

American Standard Safety Code for Protection of Heads, Eyes, and Respiratory Organs (ASAZ 2.1).

[5] Available from the American Society for Testing and Materials, 1916 Race Street, Philadelphia 3, Pa.

[6] Available from the American Concrete Institute, P.O. Box 4754, Redford Station, Detroit 19, Mich.

[7] Available from American Standard Association, 10 E. 40th Street, New York 16, N.Y.

## 8. Department of the Army Technical Manuals[8]

| TM 5–610 | Maintenance and Repair, Buildings and Structures, Preventive Maintenance, Safety Requirements. |
| TM 5–616 | Carpentry (when published). |
| TM 5–617 | Roofing. |
| TM 5–618 | Painting. |
| TM 5–619 | Plumbing. |
| TM 5–620 | Buildings and Structures; Calking and Glazing. |
| TM 5–621 | Buildings and Structures; Lathing and Plastering. |
| TM 5–622 | Wharves, Shore Structures, and Dredging. |
| TM 5–623 | Reconditioning Floors and Laying Floor Coverings (when published). |
| TM 5–624 | Roads, Runways, and Miscellaneous Pavements. |
| TM 5–625 | Sheet Metal (when published). |
| TM 5–626 | Thermal Insulation of Buildings (when published). |
| TM 5–630 | Grounds Maintenance and Land Management. |
| TM 5–632 | Insect and Rodent Control. |

## 9. Other Publications

Architects' and Builders' Handbook, Kidder-Nolan, John Wiley and Sons, Inc., New York, N.Y.

Sweets File, Architectural, Sweets Catalogue Service, 119 West 40th Street, New York, N.Y.

Brick Engineering Handbook, Structural Clay Products Institute, 1756 K Street, N.W., Washington, D.C.

Facts about Concrete Masonry, National Concrete Masonry Association, 33 West Grand Avenue, Chicago, Ill.

Building Estimator's Reference Book, Frank B. Walker, Chicago, Illinois.

Manual of Accidents in Construction, Associated General Contractors of America, Munsey Building, Washington, D.C.

[8] Available from the Adjutant General, Department of the Army, ATTN: AGAM, Washington, D.C.

AGO 10179A

# APPENDIX II

## SPECIFICATION FOR CONCRETE (FOR BUILDING CONSTRUCTION)

### 1. Scope

This section covers concrete work, complete.

### 2. Work Not Included

This section does not include any concrete for the following work: ...........................................

--------------------------------------------------

### 3. Applicable Publications

The following publications of the issues listed below, but referred to thereafter by basic designation only, form a part of this specification to the extent indicated by the references thereto:

*a. Federal Specifications.*

| | |
|---|---|
| L–S–137 & Am–i | Screening, Plastic Coated Fibrous Glass, Insect. |
| O–C–00105 (GSA–FSS) | Calcium Chloride, Dihydrate and Calcium Chloride, Anhydrous; Technical. |
| R–T–143 & Am–1 | Tars; (for use in) Road Construction. |
| HH–F–191a & Am–2 | Felt; Asphalt-Saturated (for) Flashings, Roofing, and Waterproofing. |
| HH–F–341a | Filler, Expansion-Joint, Preformed, Nonextruding and Resilient-Types (for Concrete). |
| HH–I–526a | Insulation Board, Thermal-Acoustical Mineral Wool (for Roofs). |
| HH–I–551a & Am–2 | Insulation Block, and Pipe-Covering, Thermal Cellular Glass. |
| HH–I–562 | Insulation, Thermal, Mineral Wool, Block or Board and Pipe Insulation (Molded Type). |
| QQ–B–101c & Am–1 | Bases, Metal; (for) Plaster and Stucco Construction. |
| QQ–S–632 | Steel Bar, Reinforcing, (for) Concrete. |
| SS–A–281b & Am–1 | Aggregate; (for) Portland-Cement-Concrete. |
| SS–C–185 & Am–1 | Cement, Natural (for Use as a Blend with Portland Cement). |
| SS–C–192f | Cements, Portland. |
| SS–C–197 | Cements, Portland, Blast-Furnace Slag. |
| SS–C–218 | Cement, Slag. |
| SS–S–159 | Sealer; Cold-Application Mastic Type, For Joints in Concrete. |
| SS–S–164 & Am–1 | Sealer; Hot-Poured Type, for Joints in Concrete. |
| SS–S–00167a (Army–CE) | Sealing Compound, Jet Fuel Resistant, Hot Applied, Concrete Paving. |
| SS–T–321 & Am–1 | Tile; Structural, Clay, Floor. |
| TT–C–598 & Am–2 | Compound, Calking; Plastic (for Masonry and Other Structures). |
| UU–P–264a | Paper, Concrete-Curing, Waterproofed (Kraft). |
| WW–P–406a | Pipe, Steel (Seamless and Welded (for Ordinary Use)) |
| AAA–S–121b & Am–1 | Scales, Weighing; General Specifications. |
| CCC–B–811 & Am–3 | Burlap; Jute. |
| LLL–F–321b & Am–1 | Fiberboard; Insulating. |
| LLL–H–35 | Hardboard, Fibrous-Felted (Fiberboard). |
| PPP–T–60 | Tape; Pressure Sensitive Adhesive, Waterproof—for Packaging and Sealing. |

*b. Federal Standard.*

No. 158   Cements, Hydraulic; Sampling, Inspection, and Testing.

*c. Military Specification.*

MIL–S–13518B  Sealer, Surface, Wood Preservative.

*d. American Association of State Highway Officials Specifications.*

M 73   Cotton Mats for Curing Concrete
M 141   Slow Curing Liquid Asphaltic Road Material

*e. American Concrete Institute Publications.*

ACI 315   Manual of Standard Practice for Detailing Reinforced Concrete Structures.
ACI 318   Building Code Requirements for Reinforced Concrete.

*f. American Society of Heating and Air-Conditioning Engineers Publication.*

Guide (Current Issue).

*g. American Society for Testing and Materials Standards.*

A 82    Cold-Drawn Steel Wire for Concrete Reinforcement.

A 185    Welded Steel Wire Fabric for Concrete Reinforcement.

C 31    Making and Curing Concrete Compression and Flexure Test Specimens in the Field.

C 39    Compressive Strength of Molded Concrete Cylinders.

C 42    Securing, Preparing, and Testing Specimens for Hardened Concrete for Compressive and Flexural Strengths.

C 78    Flexural Strength of Concrete (Using Simple Beam With Third-Point Loading).

C 90    Hollow Load-Bearing Concrete Masonry Units.

C 94    Ready-Mixed Concrete.

C 156    Water Retention Efficiency of Liquid Membrane-Forming Compounds and Impermeable Sheet Materials for Curing Concrete.

C 165    Compressive Strength of Preformed Block-Type Thermal Insulation.

C 192    Making and Curing Concrete Compression and Flexure Test Specimens in the Laboratory.

C 231    Air Content of Freshly Mixed Concrete by the Pressure Method.

C 272    Water Absorption of Core Materials for Structural Sandwich Construction.

C 330    Lightweight Aggregates for Structural Concrete.

C 332    Lightweight Aggregates for Insulating Concrete.

C 360    Ball Penetration in Fresh Portland Cement Concrete.

D 88    Test for Viscosity by Means of the Saybolt Viscosimeter.

D 92    Test for Flash and Fire Points by means of Cleveland Open Cup.

D 155    Test for Color of Lubricating Oil and Petrolatum by Means of ASTM Union Colorimeter.

D 882    Tensile Properties of Thin Plastic Sheets and Films.

E 96    Measuring Water Vapor Transmission of Materials in Sheet Form.

## 4. General

Full cooperation shall be given other trades to install embedded items. Suitable templates or instructions, or both, will be provided for setting items not placed in the forms. Embedded items shall have been inspected, and tests for concrete or other materials or for mechanical operations shall have been completed and approved, before concrete is placed.

## 5. Materials

The following materials shall conform to the respective specifications and other requirements stipulated below.

*a. Abrasive Aggregate.* Abrasive aggregate shall consist of not less than 55 percent aluminum oxide or silicon carbide abrasive ceramically bonded together to form a homogeneous material sufficiently porous to provide a good bond with portland cement. The aggregate shall be unaffected by freezing, moisture or by cleaning compounds. The aggregate shall be well graded in size from particles retained on No. 50 sieve to particles passing No. 8 sieve, and shall have an abrasive hardness of not less than 40 as determined by the test for wear resistance in the National Bureau of Standards Report BMS 98. The color of the aggregate shall be black or gray as selected by the Contracting Officer.

*b. Aggregate (Except for Lightweight Concrete).* Federal Specification SS-A-281, class 1 and class 2. Coarse aggregate shall be well graded from fine to coarse within the prescribed limits. The nominal sizes shall be _____ inches to No. 4 for class A concrete; _____ inches to No. 4 for class B concrete; _____ inches to No. 4 for class C concrete; and ¾ inch to No. 4 for class AA concrete. For class P concrete the aggregate shall be _____ inches in maximum size and shall be graded within the following limits:

| Sieve designation (US Standard square mesh) | Percentages passing sieves | | |
|---|---|---|---|
| | 2½-inch maximum | 2-inch maximum | 1½-inch maximum |
| 3-inch | 100 | | |
| 2½-inch | 93–100 | 100 | |
| 2-inch | 65–85 | 93–100 | 100 |
| 1½-inch | 45–60 | 75–90 | 93–100 |
| 1-inch | | 50–70 | 65–80 |
| ¾-inch | 20–35 | 35–55 | 45–65 |
| ⅜-inch | 5–15 | 10–20 | 15–30 |
| No. 4 | 0–5 | 0–5 | 0–5 |

*c. Aggregate (for Lightweight Concrete).* Aggregate for class D concrete (lightweight structural) shall conform to ASTM Standard C 330. Aggregate for class E concrete (lightweight fill, other than insulating) and class F concrete (lightweight insulating) shall conform to ASTM Standard C 332. Lightweight aggregates shall have the porosity, strength, weight, and gradation required to produce concrete having the characteristics specified hereinafter.

*d. Anchorage Items.* Slots and inserts for anchoring masonry and mechanical items to concrete, and clips for anchoring wood sleepers to concrete, shall be of standard manufacture and of types required to engage with the anchors to be provided and installed therein under other sections of these specifications, and shall be subject to approval.

(1) *Slots* shall be dovetail-type, formed of not lighter than 24-gage galvanized sheet steel, and shall be furnished with felt or fiber fillers.

(2) *Inserts for suspended ceilings.* Wire inserts for attachment of wire hangers for suspended ceilings shall be not lighter than 7-gage galvanized steel wire. When the use of flat steel hangers is approved, inserts of the same section shall be set in the concrete.

(3) *Inserts for shelf angles and bolt hangers* shall be of iron and of sturdy design having adequate strength for the load to be carried. Inserts for shelf angles shall have an integral loop at the back, shall be slotted to receive a special-headed bolt, shall have provision for nonslip vertical adjustment of shelf angle, and unless otherwise indicated, shall be furnished complete with special-headed bolt not smaller than $5/8$ inch in diameter and of the required length and fitted with hexagonal nut. Inserts for bolt hangers shall be either threaded or slotted as required by the type of hanger to be used. Threaded inserts shall have integral lugs to prevent turning.

(4) *Sleeper clips* for anchoring wood sleepers to concrete slabs shall be formed of not lighter than 20-gage galvanized sheet steel and shall have legs providing embedment of not less than $1\frac{3}{4}$ inches in the concrete, and a pair of wings, each punched with two holes for attachment to sleepers.

*e. Asphalt for Asphalt-Concrete Fill.* AASHO Specification M 141, grade SC-3, SC-4, or SC-5.

*f. Asphalt Saturated Felt.* Federal Specification HH-F-191, 30-pound.

*g. Cement.* Only one brand of any one type of cement shall be used for exposed concrete surfaces of any individual structure. Cement reclaimed from cleaning bags or leaking containers shall not be used. Cement shall be used in the sequence of receipt of shipments, unless otherwise directed.

(1) *Portland cement.* Federal Specification SS-C-192, type I or type II (type I-A or type II-A).

(2) *High-early-strength portland cement.* Federal Specification SS-C-192, type III (type III-A).

(3) *Portland blast-furnace-slag cement.* Federal Specification SS-C-197, type I S (type I SA).

(4) *Slag-cement.* Federal Specification SS-C-218, type S (type SA).

(5) *Natural cement.* Federal Specification SS-C-185, type N (type NA).

(6) *Air-entraining cement.* Federal Specification SS-C-192, type I-A, II-A, III-A, or I SA. When a blend of natural cement and portland cement is used, the air-entraining agent shall be interground in the natural cement only. When a blend of portland cement and slag cement is used, the air-entraining agent shall be interground in the portland cement only.

*h. Coloring for Floors.* Finely ground, unfading mineral oxides prepared especially for the purpose and interground with the cement.

*i. Curing materials.*

(1) *Waterproof paper.* Federal Specification UU-P-264.

(2) *Mats.* AASHO M-73.

(3) *Burlap.* Federal Specification CCC-B-811.

(4) *Polyethylene sheeting* shall be white, free of visible defects, uniform in appearnace, and not less than 0.004 inch thick unless otherwise specified, and shall pass the following tests. Certified copies of these test results shall be furnished the Contracting Officer.

(a) *Water-vapor permeance.* ASTM Standard Method E 96—permeance not to exceed 0-5 perm.

(b) *Tensile strength and elongation.* ASTM Standard Method D 882; Methods A or C—minimum requirements as follows:

| | Tensile strength | |
| --- | --- | --- |
| | Machine direction | Transverse direction |
| 0.004 inch thick | Method A, 1700 pounds | 1200 pounds |
| | Method C, 2300 pounds | 1650 pounds |

| | Elongation | |
| | *Machine direction* | *Transverse direction* |
| --- | --- | --- |
| Method A, 250% | | 350% |
| Method C, 475% | | 575% |

| | Tensile strength | |
| | *Machine direction* | *Transverse direction* |
| --- | --- | --- |
| 0.006 inch thick | Method A, 1700 pounds | 1200 pounds |
| | Method C, 2060 pounds | 1625 pounds |

| | Elongation | |
| | *Machine direction* | *Transverse direction* |
| --- | --- | --- |
| Method A, 250% | | 350% |
| Method C, 575% | | 610% |

(c) *Moisture retention.* ASTM Tentative Method C 156.

(d) *Fungus resistance.* Material shall show no fungus growth when tested for mildew resistance in accordance with the method set forth in Federal Specification L–S–137.

(5) *Polyethylene-coated waterproof paper.* The waterproof paper shall conform to Federal Specification UU–P–264. The polyethylene coating shall be clear, shall have a minimum thickness of 0.002 inch, and shall be permanently bonded to the waterproof paper. The polyethylene-coated paper shall conform to the water-vapor-permeance, moisture-retention, and fungus-resistance requirements stipulated hereinbefore for polyethylene sheeting.

*j. Drainage Fill.* Drainage fill under concrete floor slabs and areaways shall consist of clean crushed rock, crushed or uncrushed gravel, or other similar approved free-draining material of such size as will pass a 1½-inch screen and not more than 5 percent will pass a No. 4 screen. Drainage fill shall contain no earth, clay, or any other foreign substances, that will be deleterious to pipe or conduit.

*k. Expansion Joints.*

(1) *Premolded expansion-joint filler strips.* Federal Specification HH–F–341, sizes indicated on the drawings.

(2) *Joint sealer, hot-poured type.* Federal Specification SS–S–164, delivered to the building site in manufacturer's sealed containers.

(3) *Joint sealer, cold-application mastic type.* Federal Specification SS–S–159.

(4) *Joint sealer, hot-poured, jet-fuel-resistant type (for hangar floors).* Federal Specification SS–S–167, delivered to the building site in manufacturer's sealed containers.

(5) *Calking compound, gun-type.* Federal Specification TT–C–598, grade 1.

*l. Filler Units.* Filler units between ribs for long-span concrete slabs: One of the following:

(1) *Structural clay tile.* Federal Specification SS–T–321, grade M, scratched or scored on all faces.

(2) *Concrete masonry units.* ASTM Standard C 90.

*m. Forms.* Wood, metal, structural hardboard, or other approved material that will not adversely affect the surface of the concrete and that will produce or facilitate obtaining the specified surface finish of the concrete.

(1) *Wood.*

(a) *Unexposed concrete surfaces.* No. 2 Common or better lumber.

(b) *Exposed concrete surfaces.* Dressed-and-matched boards uniformly thick and not more than 10 inches wide.

(c) *Rubbed or smooth surfaces.* Plywood or forms with linings as specified below.

(2) *Plywood.* Commercial-Standard Douglas-fir, moisture-resistant, concrete-form plywood, not less than 5-ply and at least $\frac{9}{16}$ inch thick.

(3) *Metal.* Metal forms of approved type that will produce surfaces equal to those specified for wood forms.

(4) *Hardboard forms.* A hard-pressed fiberboard conforming to Federal Specification LLL–H–35, especially treated for concrete-form use, not less than ¼-inch thick.

(5) *Form lining.*

(a) *Plywood.* Commercial-Standard Douglas-fir, concrete-form exterior, 3-ply, not less than ¼-inch thick.

(b) *Fiberboard.* A treated hard-pressed fiberboard, Federal Specification LLL–H–35, type I, class 2, not less than $\frac{3}{16}$-inch thick, with one smooth side.

(c) *Absorptive-type lining.* Material shall have an absorption coefficient sufficient to eliminate voids and pitting and to produce a dense and uniform concrete surface, shall not interfere with the normal chemical reaction of or discolor the cement, shall be easily cut for fitting and easy to remove at the end of the curing period.

*n. Form Oil.* An approved colorless mineral oil, not darker than ASTM No. 3 in accordance with ASTM D–155, free of kerosene, with a viscosity of not less than 70 seconds nor more than 110 seconds (Saybolt Universal) at 100° F., except that when used on hardboard forms, the viscosity shall be not less than 250 seconds at 100° F. Flash point shall be not less than 300° F. (open cup). Viscosity and flash point shall be determined in accordance with ASTM Standards D–88 and D–92, respectively.

*o. Form Sealer.* Military Specification MIL–S–13518.

*p. Form Ties.* An approved design, fixed or adjustable in length, free of devices that will leave a hole larger than ⅞-inch in diameter in surface of concrete, and when used where discoloration of the concrete would be objectionable, the metal remaining after the removal of the external parts shall be not less than 1 inch below the finished surface.

*q. Perimeter Insulation.* Rigid board type not affected by moisture, vermin, or fungi, and shall be one of the following:

(1) *Cellular glass.* Federal Specification HH–I–551, type 1.

(2) *Insulation board.* Federal Specification HH–I–526, with waterproof facing on one face and all edges.

(3) *Cellular plastic* shall be an expanded polystyrene meeting the following requirements:

(a) *Conductivity* shall not exceed 0.38 B.t.u. per square foot per hour per degree F. per inch of thickness at 70° F.

(b) *Compressive strength* shall be not less than 8 psi when the board is compressed to a deformation of 5 percent of its original thickness when tested under ASTM Standard C–165, modified to change drying temperature to 150° F.

(c) *Moisture absorption* shall not exceed 0.05 percent by volume when tested by the immersion method in accordance with ASTM Standard C–272, using drying oven at 122° F. plus or minus 5.4°.

*r. Reinforcement.*

(1) *Bars.* Federal Specification QQ–S–632, type II, grade C, D, E, or G except as otherwise indicated.

(2) *Column spirals.* Plain cold-drawn wire conforming to ASTM Standard A 82 or hot-rolled rods.

(3) *Dowels for load transfer in floors.* Of type, design, weight, and dimensions shown, Dowel bars shall be type I steel bars conforming to Federal Specification QQ–S–632. Dowel pipe shall be steel pipe conforming

to Federal Specification WW–P–406, weight B, class 1.

(4) *Mesh reinforcement.* ASTM Standard A 185, except as otherwise specified hereinafter. When indicated in slabs on fill, mesh shall be of the sizes indicated. Mesh in applied floors, roof fills, and concrete subfloor fills on wood construction shall be of the size indicated and in no case larger than 4- by 4-inch of not lighter than 10-gage wire. Mesh wrapping for structural steel members shall be 4- by 4-inch mesh of 11-gage galvanized wire. Mesh for temperature reinforcement in fills over membrane waterproofing, over insulation in floors of refrigerated spaces, and over lead X-ray protection shall be 1½- or 2-inch hexagonal mesh galvanized fencing of not lighter than 19-gage wire.

(5) *Metal floor lath for use over steel joists to receive concrete floor slabs.* Expanded metal lath conforming to Federal Specification QQ–B–101, type F3/8R, 4 pounds per square yard, for joist spacing up to and including 24 inches, and type F3/4R, 0.06 pound per square foot, for joist spacing over 24 and up to and including 30 inches; or 3- by 4-inch mesh of 12-gage cold-drawn electrically welded galvanized wire with a heavy water-resistant paper backing.

(6) *Mill reports.* Certified copies of mill reports shall accompany deliveries of reinforcing steel, except mesh reinforcement and metal floor lath, on projects using 15 tons or more, except that if grade G bars are furnished a certificate shall be furnished stating that the bars have been manufactured in accordance with the requirements of the Rail Steel Bar Association.

*s. Tar for Tar-Concrete Fill.* Federal Specification R–T–143, grade RT–7, RT–8, or RT–9.

*t. Vapor Barrier.* Vapor barrier for use under slabs on grade shall be fungi-resistant when tested by the method set forth in Federal Specification L–S–137 and shall have a vapor permeance rating not exceeding 0.5 perm as determined by method set forth in ASTM E–96, procedure E. Vapor barrier shall be asphalt-saturated waterproof reinforced kraft paper, clear polyethylene sheeting 0.006 inch thick, polyethylene-coated asphalt-saturated reinforced kraft paper, two layers of 30-pound asphalt-saturated felt solid-mopped with hot bitumen, or other similar material meeting above requirements for fungi resistance and vapor permeance.

*u. Joint Filler.* Ventilating and expansion joint filler for lightweight insulating concrete shall be either of the following:

  (1) *Glass-fiber board.* Federal Specification HH-I-526.
  (2) *Mineral-wool board.* Federal Specification HH-I-562, type I, class 1.

*v. Water.* Clean, fresh, and free from injurious amounts of mineral and organic substances.

## 6. Admixtures

*a. Air-entraining agents* shall be selected well in advance of concrete placing, and the contractor shall provide facilities satisfactory to the Contracting Officer for ready procurement of adequate test samples. The use of air-entraining cement produced at the mill or an air-entraining agent added at the mixer shall be optional with the contractor. Air-entraining concrete shall have an air content by volume, 4.5 percent plus or minus 1.5 percent, as derermined by ASTM Standard C 231. When necessary to increase the air content, additional air-entraining admixture identical with that already in the mixture shall be added at the mixer. Tests for determining air content shall be the responsibility of the contractor, subject to inspection and approval.

*b. Accelerating agent* shall be calcium chloride conforming to Federal Specification O-C-105, type I or type II. Use and amount of material shall be subject to approval of the Contracting Officer.

*c. Other admixtures,* except air-entraining and accelerating agents, shall be used only on written approval.

*d. Tests* of admixtures will be made by the Government in accordance with applicable Federal or ASTM specifications or as otherwise prescribed.

## 7. Samples and Testing

*a. Testing* of the aggregate and reinforcement shall be the responsibility of the contractor. The testing agency shall be approved. Testing of end items is the responsibility of the Government. Samples of concrete for strength tests of end items shall be provided and stored by the contractor when and as directed.

*b. Cement* shall be tested as prescribed in the applicable referenced specification under which it is furnished. The soundness or autoclave expansion test for a blended cement shall be as specified in Federal Standard 158, and the maximum percentage of expansion permitted in the blended product shall not exceed the limit specified for the portland cement used in the blend.

Cement

*For projects requiring 1,200 bags or less*

(may be accepted on the basis of mill tests and the manufacturer's certification of compliance with the specification, provided the cement is the product of a mill with a record for the production of high-quality cement for the past 3 years. Certificates of compliance for each mill lot of cement furnished from different mills in mixed shipment and for each separate shipment from the same mill, shall be furnished by the contractor prior to the use of the cement in the work. This requirement is applicable to cement for job-mixed, ready-mixed, or transit-mixed concrete. When no certificate of compliance is furnished or when, in the opinion of the Contratcing Officer, the cement furnished under certificate of compliance may have become damaged in transit, or deteriorated because of age or improper storage, the cement proposed for use will be sampled at the mixing site by representatives of the Government and tested for conformance to the specification at no expense to the contractor. Access to the cement and facilities for sampling shall be readily afforded the Government's agent. Cement being tested shall not be used in the work prior to receipt by the contractor of written notification from the Contracting Officer that the cement has satisfactorily passed the 7-day tests. In the case of job-mixed concrete, cement failing to meet test requirements shall be removed from the site. In the case of ready-mixed and transit-mixed concrete, cement at batching plants failing to meet test requirements shall not be used ir Government work.)

*For projects requiring more than 1,200 bags*

(shall be sampled either at the mill or at the site of the work. Tests will be made by or under the supervision of the Contracting Officer at the expense of the Government. Cement shall be tested by an approved testing laboratory or testing agency with test data subject to verification by the National Bureau of Standards, or by Government laboratories properly equipped for performance of the required tests. No cement shall be used until notice has been given by the Contracting Officer that the test results are satisfactory. Cement that has been stored, other than in the bins at the mills, for more than 4 months after being tested shall be retested before use. Cement delivered at the site of the work and later found under test to be unsuitable shall be removed from the work and its vicinity.)

c. *Aggregate* shall be tested as prescribed in Federal Specification SS–A–281.

d. *Reinforcing* bars shall be tested as prescribed in Federal Specification QQ–S–632. Mesh reinforcement shall be tested as prescribed in ASTM Standard A 185.

e. *Concrete* (except lightweight insulating:)

(1) *Strength tests during the work.* The contractor shall provide for test purposes (three sets of three cylinders and three sets of three beams) taken from each 250 cubic yards or fraction thereof, or each day's pour, whichever is less, of each class of concrete placed. Test specimens shall be made and cured in accordance with ASTM Standard C 31. Specimens shall be cured under laboratory conditions except that the Contracting Officer may require curing under field conditions when he considers that there is a possibility of the air temperature's falling below 40° F. (Cylinders shall be tested in accordance with ASTM Standard C 39.) (Beams shall be tested in accordance with ASTM Standard C 78.) The test result shall be the average of the strengths of the test specimens except that if one specimen in a set of three shows manifest evidence of improper sampling, molding, or testing, the test result shall be based on the average of the remaining two specimens. If two specimens in a set of three show such defects, the results of the set shall be discarded and average strength determined from test results of the other two sets. The standard age of test shall be 28 days, but 7-day tests may be used, with the permission of the Contracting Officer, provided that the relation between the 7-day and 28-day strengths of the concrete is established by tests for the materials and proportions used. If the average of the strength tests of the specimens cured under laboratory controls, for any portion of the work, falls below the minimum allowable (compressive or flexural) strength at 28 days required for the class of concrete used in that portion, the Contracting Officer shall have the right to order a change in the proportions or the water content of the concrete, or both, for the remaining portions of the work at the contractor's

expense. If the average strength of the specimens cured under actual field conditions falls below the minimum allowable strength, the Contracting Officer may require changes in the conditions of temperature and moisture necessary to secure the required strength.

(2) *Tests of hardened concrete in, or removed from, the structure.* Where the results of the strength tests of the control specimens indicate the concrete as placed does not meet specification requirements or where there is other evidence that the quality of the concrete is below specification requirements, one or both of the following tests will be required:

(a) Core-boring tests conforming to ASTM Standard C–42.

(b) Load tests in accordance with Section 202 of the ACI Building Code (ACI 318).

(3) *Core-boring or load test.* Where the core-boring-test results indicate that the in-place concrete does not meet specification requirements, the load test shall be made. In the event the load test indicates that the concrete placed does not conform to the drawings and specifications, the cost of both tests shall be borne by the contractor and measures as prescribed by the Contracting Officer shall be taken to correct the deficiency at no additional expense to the Government. If the boring test, or the load test, if required, indicate that the concrete as placed conforms to the drawing and specification requirements, the cost of the test will be borne by the Government.

## 8. Storage

Storage accommodations for concrete materials shall be subject to approval and shall afford easy access for inspection and identification of each shipment in accordance with test reports.

a. *Cement.* Immediately upon receipt at site of work, cement shall be stored in a dry, weathertight, properly ventilated structure, with adequate provision for prevention of absorption of moisture.

b. *Aggregate.* Storage piles of aggregate shall be so located as to assure good drainage, to preclude inclusion of foreign matter, and to preserve the gradation. Sufficient live storage shall be maintained to permit segregation of shipments from different sources, and to assure placement of concrete at the required rate. Storage of aggregate for cold-weather placement shall permit access for required heating operations.

## 9. Forms

Forms complete with centering cores, and molds, shall be constructed to conform to shape, form, line, and grade required, and shall be maintained sufficiently rigid to prevent deformation under load. Studs shall be spaced sufficiently close to prevent deflection of form material and consequent waviness in surface of concrete.

*a. Design.* Joints shall be sufficiently tight to prevent leakage of grout during placing and shall be arranged vertically or horizontally to conform to the pattern of the design. Forms placed on successive units for continuous surfaces shall be fitted to accurate alinement to assure a smooth completed surface free from irregularities. In long spans, where intermediate supports are not possible, the anticipated deflection in the forms due to weight of fresh concrete shall be accurately figured and taken into account in the design of the forms, so that finished concrete members will have true surfaces conforming accurately to desired lines, planes, and elevations. If adequate foundation for shores cannot be secured, trussed supports shall be provided. Temporary openings shall be arranged in wall and column forms and where otherwise required, to facilitate cleaning and inspection. Lumber once used in forms shall have nails withdrawn and surfaces to be exposed to concrete carefully cleaned before reuse. Forms shall be readily removable without hammering or prying against the concrete.

*b. Ties.* Form ties shall be of suitable design and adequate strength for the purpose. Wire ties will not be permitted where discoloration of the finished surface would be objectionable. Bolts and rods that are to be completely withdrawn shall be coated with grease.

*c. Joints.* Corners of columns, girders, beams, foundation walls projecting beyond overlying masonry, and other exposed joints in more than one plane, unless otherwise indicated or directed, shall be beveled, rounded, or chamfered by moldings placed in the forms.

*d. Coating.* Forms for exposed surfaces, except those with absorptive lining, shall be coated with oil before reinforcement is placed. Surplus oil on form surfaces and any oil on reinforcing steel shall be removed. Forms for surfaces not exposed to view may be thoroughly wet with water, in lieu of oiling, immediately before placing of concrete, except that in cold weather with probable freezing temperatures,

oiling shall be mandatory. Wood forms for concrete that is to be painted shall be coated with sealer instead of with oil or water.

*e. Removal.* Forms shall be removed only with approval of the Contracting Officer, in a manner to insure complete safety of the structure after the following conditions have been met. Where the structure as a whole is supported on shores, the forms for the beam and girder sides, for the columns, and for similar vertical structural members may be removed after 24 hours, provided concrete is sufficiently hard not to be injured thereby. Supporting forms or shoring shall not be removed until structural members have acquired sufficient strength to support safely their own weight and any construction and/or storage load to which they may be subjected, but in no case shall they be removed in less than 6 days, nor shall forms used for curing be removed before expiration of curing period except as provided hereinafter under paragraph CURING. Care shall be taken to avoid spalling the concrete surface. Wood forms shall be completely removed from under steps and similar spaces, through temporary openings if necessary.

(1) *Control tests.* Results of suitable control tests will be used as evidence that concrete has attained sufficient strength to permit removal of supporting forms. Cylinders required for control tests shall be provided in addition to those otherwise required by this specification. Test specimens shall be removed from molds at end of 24 hours and stored in the structure as near points of sampling as possible, shall receive insofar as practicable the same protection from the elements during curing as is given those portions of the structure which they represent, and shall not be removed from the structure for transmittal to the laboratory prior to expiration of three-fourths of the proposed period before removal of forms. In general, supporting forms or shoring shall not be removed until strength of control-test specimens has attained a value of at least 1,500 pounds for columns and 2,000 pounds for all other work. Care shall be exercised to assure that the newly unsupported portions of the structure are not subjected to heavy construction or material loading.

(2) *Clamps.* Tie-rod clamps to be entirely removed from the wall shall be loosened 24 hours after concrete is placed, and form

ties, except for a sufficient number to hold forms in place, may be removed at that time. Ties wholly withdrawn from wall shall be pulled toward inside face.

(3) *Filling tie-rod or bolt holes.* Holes left by bolts or tie rods shall be filled solid within 12 hours after removal of forms, with cement mortar blended to match adjacent surface. Holes passing entirely through wall shall be filled from inside face with a device that will force the mortar through to outside face, using a stop held at the outside wall surface to insure complete filling. Holes which do not pass entirely through walls shall be packed full. Excess mortar at face of filled holes shall be struck off flush.

## 10. Reinforcing Steel

Reinforcing steel fabricated to shapes and dimensions shown, shall be placed where indicated on drawings or where required to carry out the intent of the drawings and specifications. Before being placed, reinforcing steel shall be thoroughly cleaned of loose or flaky rust, mill scale, or coating, including ice, and of any other substance that would reduce or destroy the bond. Reinforcing steel reduced in section shall not be used. After any substantial delay in the work, previously placed reinforcing steel left for future bonding shall be inspected and cleaned. Reinforcing steel shall not be bent or straightened in a manner injurious to the steel. Bars with kinks or bends not shown on drawings shall not be placed. The use of heat to bend or straighten reinforcing steel will be permitted only if the entire operation is approved. In slabs, beams, and girders, reinforcing steel shall not be spliced at points of maximum stress. Laps or splices shall be of adequate length to transmit stresses and, unless otherwise indicated, shall conform to the table in ACI 315. Splices in adjacent bars shall be staggered. Splices in columns, piers, and struts shall be lapped sufficiently to transfer the full stress by bond.

*a. Design and Details.* Unless otherwise indicated the design of reinforced concrete structures shall conform to ACI 318, and the details of reinforcing steel shall conform to ACI 315. Unless otherwise indicated, construction shall conform to the following requirements:

(1) *Concrete covering over steel reinforcement.* The thickness of the concrete covering over steel reinforcement shall be not less than the diameter of round bars, or less than 1½ times the side dimensions of square bars, and in the following specific instances not less than specified below:

Footings and other principal structural members in which concrete is deposited against the gound. — 3 inches between steel and ground.

Where concrete surfaces, after removal of forms, are exposed to weather or ground:

For bars more than ⅝ inch in diameter. — 2 inches.

For bars ⅝ inch or less in diameter. — 1½ inches.

Where surfaces are not directly exposed to weather or ground:

For slabs and walls_____ ¾ inch.

For beams, girders, and tied columns. — 1½ inches.

For spiral columns (covering to be cast monolithically with core). — 1½ inches or 1½ times maximum size of aggregate.

For concrete-joist floors in which clear distance between joists is not more than 30 inches. — ¾ inch.

Exposed reinforcement bars intended for bonding with future extensions shall be protected from corrosion by adequate covering.

(2) *Spiral reinforcement* shall consist of evenly spaced continuous spirals held firmly in place and true to line by vertical spacers, using at least two for spirals 20 inches or less in diameter, three for spirals 20 to 30 inches in diameter, and four for spirals more than 30 inches in diameter or composed of spiral rods ⅝ inch or larger in size. The spirals shall be of such size and so assembled as to permit handling and placing without being distorted from the designed dimensions. Anchorage of spiral reinforcement shall be provided by 1½ extra turns of spiral rod or wire at each end of the spiral unit. Splices, when necessary, shall be made in spiral rod or wire by welding or by a lap of 1½ turns. The reinforcing spiral shall extend from the floor level in any story or from the top of the footing in the basement, to the level of the lowest horizontal reinforcement in the overhead slab, drop panel, beam, or girder. In a column with a capital, the spiral shall

extend to a plane at which the diameter or width of the capital is twice that of the column.

(3) *Steel in walls and lintel beams*, unless otherwise shown, shall be continuous throughout the length of the various members. Splices shall not occur at critical sections.

(4) *Stirrup spacer bars:* All stirrups, except ties, shall be held in place by two ⅜-inch spacer bars extending the full length of the portion of the beam or girder occupied by stirrups.

(5) *Outside bars* of slab reinforcement, both main and temperature, parallel to beams, girders, or walls, shall be placed not over one-half bar spacing from the adjacent face of each member.

(6) *Wire-mesh reinforcement between expansion joints in slabs* shall be continuous and shall have joints lapped at least one full mesh. Lapping of sheets shall be staggered to avoid continuous lap in either direction. Reinforcement shall be supported by standard accessories for slabs above grade and by properly sized precast concrete blocks for slabs on earth.

(7) *Fully encased structural steel members* shall be wrapped with 4- by 4-inch mesh of 11-gage galvanized wire applied around the steel over spacers to provide ¾-inch clearance from the metal. The edges of the mesh shall be lapped and tied and shall have loose ends made fast with not lighter than 16-gage galvanized wire. Beam clips or caging of an approved type, with spacing of members not exceeding 6 inches in both directions, may be substituted for the wire mesh specified provided such material has an equal cross-sectional area and equivalent distribution of metal in both directions.

(8) *Shop drawings.* Shop detail and placing drawings for all reinforcing steel shall be furnished for approval.

*b. Supports.* With the exception of temperature reinforcement, which shall be tied to main steel approximately 24 inches on centers, reinforcement shall be accurately placed and securely tied at all intersections and splices with 18-gage black annealed wire, and shall be securely held in position during the placing of concrete by spacers, chairs, or other approved supports. Wire-tie ends shall point away from the form. Unless otherwise indicated, the number, type, and spacing of supports shall conform to the ACI 315.

*c. For Slabs on Grade (Over Earth or Over Drainage Fill) and for Footing Reinforcement.* Bars or mesh shall be supported on precast concrete blocks, spaced at intervals required by size of reinforcement used, to keep reinforcement the minimum height specified above the under side of slab or footing.

## 11. Classes of Concrete and Usage

*a. Strength Requirements.* Concrete of the various classes required shall be proportioned and mixed for the following strengths:

| Class | Minimum allowable compressive strength at 28 days[1] | Class | Minimum allowable compressive strength at 28 days[1] | Minimum allowable flexural strength at 28 days[1] |
|---|---|---|---|---|
| | Pounds per square inch | | Pounds per square inch | Pounds per square inch |
| AA | 3750 | D | 2000 | --- |
| A | 3000 | E | 1000 | --- |
| B | 2500 | F | 125 | --- |
| C | 2000 | P | --- | (------) |

[1] Concrete made with high-early-strength cement shall have a 7-day strength equal to the specified 28-day strength for concrete of the class specified made with ordinary portland cement.

*b. Usage.* Concrete of the various classes shall be used as follows:

(1) *Class AA concrete.* For containers for liquids and for such other work as indicated.

(2) *Class A concrete.* For_____ and for such other reinforced work as indicated.

(3) *Class B concrete.* For all reinforced work not otherwise indicated.

(4) *Class C concrete.* For all concrete not reinforced except as otherwise indicated.

(5) *Class D concrete (lightweight structural).* For_____and as otherwise indicated.

(6) *Class E concrete (lightweight fill, other than insulating).* For_____and as otherwise indicated.

(7) *Class F concrete (lightweight insulating).* For insulating roof fill where so indicated.

(8) *Class P concrete.* For_____ and as otherwise indicated.

## 12. Proportioning of Concrete Mixes

Concrete (except lightweight concrete) shall be proportioned by weight.

*a. Measurements.*

(1) *Cement.* A one-cubic-foot bag of portland

cement or portland blast-furnace-slag cement will be considered as 94 pounds in weight, a one-cubic-foot bag of slag cement as 82 pounds, and a one-cubic-foot bag of natural cement as 85 pounds. In determining the approved mix, portland cement may be used alone as the cementitious material, or natural or slag cement may be used together with portland cement as a blend, in a proportion not exceeding one bag of natural or slag cement to four bags of portland cement.

(2) *Water.* One gallon of water will be considered as 8.33 pounds.

(3) *Aggregate.* Fine and coarse aggregate, except for lightweight concrete, shall be measured by weight in accordance with Federal Specification SS-A-281. Coarse aggregate shall be used in the greatest amount consistent with required workability.

*b. Corrective Additions.* Corrective additions to remedy deficiencies in aggregate gradations shall be used only with the written approval of the Contracting Officer. When such additions are permitted, the material shall be measured separately for each batch of concrete.

*c. Control.*

(1) *Determination of maximum water content allowable.* The strength quality of the concrete proposed for use shall be established by tests made in advance of the beginning of operations, using the consistencies suitable for the work. Trial design batches and testing shall be the responsibility of the contractor. Specimens shall be made and cured in accordance with ASTM Standard C 192 and tested in accordance with ASTM Standard C 39 or C 78, as applicable. Curves representing relation between the water content and the average 28-day compressive or flexural strength, or earlier strength at which the concrete is to receive its full working load, shall be established for a range of values including the compressive and flexural strengths called for on the drawings. Curves shall be established by at least three points, each point representing average values from at least four test specimens. The maximum allowable water content for the concrete for the structure shall be as determined from these curves and shall correspond to a compressive strength 15 percent greater than that indicated on the drawings or to a flexural strength 5 percent greater than that indicated on the drawings. Prior to commencing operations, the contractor shall furnish a statement to the Contracting Officer giving the proportions by weight (dry) of cement and of fine and coarse aggregates that will be used in the manufacture of each class of concrete proposed for use. The statement shall be accompanied by test reports or other evidence satisfactory to the Contracting Officer attesting that the proportions thus selected will produce concrete of the qualities specified. No substitutions shall be made in the materials used in the work without additional tests in accordance herewith to show that the quality of the concrete is satisfactory.

(2) *Slump test.* In the field, consistency shall be determined in accordance with CRD-C 5. The slump shall fall within the following limits, provided the required strength is obtained:

| Type of structure | Slump in inches for vibrated concrete | |
|---|---|---|
| | Minimum | Maximum |
| General building construction[1] | 2 | 3 |
| Thin reinforced walls[1] | 3 | 4 |
| Heavy-duty floor and slab construction (Class P concrete) | 1 | 2 |

[1] The slump for nonvibrated concrete when approved by the Contracting Officer shall be from 3 to 6 inches.

## 13. Ready-Mixed Concrete

Where ready-mixed concrete is proposed for use, the mixing and transporting equipment and the method of placement shall be subject to approval. Except for materials herein specified, ready-mixed concrete shall conform to ASTM Standard C 94, except that mixing time for mixers over 1 cubic yard capacity shall be increased 15 seconds for each additional ½ cubic yard or fraction thereof of material mixed.

## 14. Batching and Mixing

*a. Type of Plant.* The batching plant and mixing equipment shall have a capacity of at least _____ cubic yards in 8 hours. Either a manual or a semi-automatic plant may be used, subject to the approval of the Contracting Officer. A manual plant is defined as one in which batch weights are set manually and materials are batched manually. A semi-automatic plant is defined as one in which batching weights are set manually, mixes are changed manually, and materials are batched automatically.

b. *Batching Plant.*

(1) *Location.* The batching plant may be located on-site or off-site.

(2) *Arrangement.* Separate bins or compartments shall be provided for fine aggregate, for the different sizes of coarse aggregate, and for bulk cements when used. The compartments shall be of ample size and so constructed that the materials will remain separate under all working conditions. In a manual plant aggregates may be weighed cumulatively in one weigh batcher on one scale; in a semiautomatic plant they may be weighed cumulatively in one weigh batcher on one scale or in separate weigh batchers with individual scales. In a semiautomatic plant, bulk cement shall be weighed on a separate scale in a separate weigh batcher. In a manual plant, bulk cement shall be weighed in a separate hopper, which may be attached to a separate scale for individual weighing, or may be attached to the aggregate hopper for cumulative weighing provided there are separate beams or dials for cement and aggregates. If cement is weighed on the same scale as the aggregates, the cement shall be weighed first and an interlock shall be provided to insure that all hoppers are empty and that the scale is in balance before the weighing of the cement is begun. Water may be measured by weight or by volume. In a semiautomatic plant, the batching controls shall be so interlocked that a new batching cycle cannot be started until all batchers are completely empty. The plant shall be so arranged as to facilitate the inspection of all operations at all times. Suitable facilities shall be provided for obtaining representative samples of concrete for uniformity tests. Delivery of materials from the batching equipment shall be within the following limits of accuracy:

Cement _____1%
Water _____1%

Aggregate _____2%
Air-entering admixture _____1%

(3) *Water-batcher and dispenser for admixture.* Equipment for batching water and the air-entraining admixture shall be provided at the batching plant or included with the mixer, as required for the type of plant used. A suitable water-measuring device shall be provided that will be capable of measuring the mixing water within the specified requirements for each batch. The mechanism for delivering water to the mixers shall be such that leakage will not occur when the valves are closed. The filling and discharge valves for the water batcher shall be so interlocked that the discharge valve cannot be opened before the filling valve is fully closed. Where the air-entraining admixture is added at the mixer, a suitable device for measuring and dispensing the admixture shall be provided. The device shall be capable of ready adjustment to permit varying the quantity of admixture to be batched. The dispenser for air-entraining admixture shall be interlocked with the batching and discharging operations of the water so that the batching and discharging of the admixture will be automatic. When use of truck mixers makes this requirement impracticable, the air-entraining admixture shall be interlocked with the sand batcher.

(4) *Moisture control.* A semiautomatic plant shall be capable of ready adjustment for the varying moisture contents of the aggregates and to change the weights of the materials being batched.

(5) *Scales.* Adequate facilities shall be provided for the accurate measurement and control of each of the materials entering each batch of concrete. The accuracy of the weighing equipment shall conform to the applicable requirements of Federal Specification AAA-S-121 for such equipment. The contractor shall provide standard test weights and any other auxiliary equipment required for the operation of each scale or other measuring device. Periodic tests shall be made in the presence of a Government inspector in such manner and at such intervals as may be directed. Upon completion of each check test and before further use of the indicating, recording, or control devices, the con-

tractor shall make such adjustments, repairs, or replacements as may be required to insure satisfactory performance. Each weighing unit shall include a visible spring-less dial that will indicate the scale load at all stages of the weighing operation, or shall include a beam scale with a beam-balancing indicator that will show the scale in balance at zero load and at any beam setting. The indicator shall have an over and under travel equal to at least 5 percent of the capacity of the beam. The weighing equipment shall be arranged so that the plant operator can conveniently observe all dials or indicators.

(6) *Recorders (for semiautomatic type)*. Not more than two accurate recorders shall be provided for a semiautomatic plant. Each recorder shall be housed in a cabinet that is capable of being locked and shall be in a position convenient for observation by the plant operator and the Government in-spector. One recorder shall produce a printed or autographic record on a single visible chart or tape of the weights of all of the aggregates as batched and after the batcher is discharged return to zero. One recorder shall produce a printed or auto-graphic record on a single visible chart or tape of the weight of the cement as batched and after the batcher has been discharged return to zero. The weight or volume of water shall likewise be recorded if batched at a central batching plant. The charts or tapes shall clearly indicate the different types of mixes used, by stamped letters, numerals, colored ink, or other suitable means; shall be so marked that variations in batch weights of each type of mix can be readily observed; shall show time of day (stamped or preprinted) at intervals of not more than 15 minutes; and shall become the property of the Government.

(7) *Protection*. All weighing, indicating, re-cording, and control equipment shall be protected against exposure to dust and weather.

c. *Concrete mixers*. Concrete mixers may be stationary mixers, truck mixers, or paving mixers of approved design. The mixers shall have a rated capacity of at least _ _ _ _ _ _ cubic feet of mixed concrete and shall not be charged in excess of the capacity recommended by the manufacturer. Mixers shall be capable of combining the materials into a uniform mixture and of discharging this mixture without segregation. Stationary and paving mixers shall be provided with an acceptable device to lock the discharge mechanism until the required mixing time has elapsed. Truck mixers shall be equipped with accurate revolution counters. The mixers or mixing plant shall include a device for automatically counting the total number of batches of concrete mixed. The mixers shall be operated at the drum speed designated by the manufacturer on the name plate. The mixing periods specified herein are based on proper control of the speed of rotation of the mixer drum, and on proper introduction of the materials into the mixer. The mixing time will be increased when such increase is necessary to secure the required uniformity and consistency of the concrete, or when test samples of concrete taken from the front, center, and back of the mixer show a difference of more than 10 percent in sand-cement or water-cement ratio. Excessive over-mixing re-quiring additions of water will not be permitted. The mixers shall be maintained in satisfactory operating condition, and mixer drums shall be kept free of hardened concrete. Mixer blades shall be replaced when worn down more than 10 percent of their depth. The use of any mixer that at any time produces unsatisfactory results shall be promptly discontinued until mixer is repaired.

(1) *Paving mixers* shall be equipped with boom and bottom-dump bucket to handle the concrete from the mixer to the form. The bucket shall be of adequate size to handle the complete batch of concrete mixed, and the boom shall be of sufficient length to permit discharge of the concrete into its final position in the form. Paving mixers may be either single-drum or dual-drum type. Dual-drum paving mixers shall be properly synchronized and the mixing time shall be determined by excluding the time required to transfer the concrete from the first to the second drum. The mixing time for each batch, after all solid materials are in the mixer drum and provided that all the mixing water is introduced before one-fourth of the mixing time has elapsed, shall be not less than 1 minute for mixers having a capacity of 1 cubic yard; for mixers of larger capacities, the minimum mixing time shall be increased 15 seconds for each additional one-half cubic yard or fraction thereof of additional concrete mixed.

Vehicles used in transporting materials from the batching plant to the mixers shall have bodies or compartments of adequate capacity to carry the materials and to deliver each batch, separate and intact, to the mixer. Except as otherwise approved by the Contracting Officer, loose cement shall be transported from the batching plant to the mixers in separate boxes or compartments equipped with windproof and rainproof covers.

(2) *Stationary mixers.* The mixing time for each batch, after all solid materials are in the mixer drum and provided that all of the mixing water is introduced before one-fourth of the mixing time has elapsed, shall be not less than 1 minute for mixers having a capacity of 1 cubic yard; for mixers of larger capacities, the minimum mixing time shall be increased 15 seconds for each additional one-half cubic yard or fraction thereof of additional concrete mixed. When a stationary mixer is used for partial mixing of the concrete (shrink-mixed) the mixing time in the stationary mixer may be reduced to the minimum necessary to intermingle the ingredients (about 30 seconds). Vehicles used for transporting central-mixed concrete shall conform to the applicable requirements of ASTM Standard C 94. Nonagitating equipment for transporting central-mixed concrete may be used when authorized in writing by the Contracting Officer. Methods and equipment for handling and depositing the concrete in the form shall be subject to the approval of the Contracting Officer.

(3) *Truck mixers* may be used when the equipment and methods are approved in writing by the Contracting Officer. Concrete so manufactured shall conform in every respect to the requirements of the specifications. When a truck mixer is used either for complete mixing (transit-mixed) or to finish the partial mixing done in a stationary mixer, each batch of concrete shall be mixed not less than 35 nor more than 75 revolutions of the drum for horizontal-discharge-type mixers and not less than 50 nor more than 100 revolutions of the drum for high-discharge-type mixers, both at the rate of rotation designated by the manufacturer of the equipment as mixing speed.

Any additional mixing shall de done at the speed designated by the manufacturer of the equipment as agitating speed. When necessary for proper control of the concrete, mixing of transit-mixed concrete will not be permitted until the truck mixer is at the site of the concrete placement.

## 15. Expansion Joints

Expansion joints shall be constructed as indicated on the drawings and as approved. In no case shall the reinforcement, corner protection angles, or other fixed metal items, embedded in or bonded into concrete, be run continuous through an expansion joint.

*a. Slab Joints.* Joints between slabs on earth and between slabs on earth and vertical surfaces, except insulated joints, shall be of premolded expansion-joint filler strips. Joints shall be one-half inch thick and the full depth of slab, unless shown otherwise on the drawings.

*b. Perimeter Insulation (Interior).* Perimeter insulation (interior) shall be installed between slabs on grade and exterior walls. Insulation shall be not less than 1 inch thick, and shall extend not less than the full depth of the slab and the drainage fill underneath the slab. Where exterior grade is below the bottom elevation of the concrete floor slab, the insulation shall extend not less than 8 inches below the elevation of the exterior grade. Insulation shall be applied to the foundation wall with either a hot or cold bituminous mastic. Joints of the insulating boards shall be sealed with the mastic used for application of the boards.

*c. Joints with Compound.* At all expansion joints in concrete floor slabs and at other joints indicated to receive joint compound, premolded expansion-joint filler strips, or other approved premolded strip material, shall be installed at the proper level below the finished floor with a slightly tapered, dressed-and-oiled wood strip temporarily secured to the top thereof. The wood strip shall be of sufficient depth to form a groove not less than $3/4$-inch deep. After the concrete has set, the wood strip shall be removed and the groove shall be filled with joint sealer, hot-poured type or cold-application mastic type. Joint grooves shall be filled approximately flush, so as to be slightly concave after drying. Where vertical expansion joints are indicated they shall be partially filled with oakum and finished off to surface with light-colored gun-type calking compound tooled slightly concave.

*d. Finish at Joints.* Edges of cement floors or

concrete slabs along expansion joints shall be neatly finished with a slightly rounded edging tool.

## 16. Construction Joints

The unit of operation shall not exceed 50 to 60 feet in any horizontal direction, unless otherwise indicated or approved by the Contracting Officer. Concrete shall be placed continuously so that the unit will be monolithic in construction. At least 48 hours shall elapse between the casting of adjoining units, unless this requirement is waived by the Contracting Officer. Construction joints, if required, shall be located near the midpoint of spans for slabs, beams, or girders, unless a beam intersects a girder at the center, in which case the joints in the girder shall be offset a distance equal to twice the width of the beam and provision for shear shall be made by use of inclined reinforcement. Joints in columns or piers shall be made at the under side of the deepest beam or girder framing thereto. Columns, piers, or walls of ordinary height shall be poured at least 2 hours before any overhead work is placed thereon. Joints not shown or specified shall be so located as to least impair strength and appearance of work. Construction joints in wall footings shall be reduced to a minimum. Except where otherwise indicated no jointing shall be made in footings or foundation work without written approval. Placement of concrete shall be at such rate that surfaces of concrete not carried to joint levels will not have attained initial set before additional concrete is placed thereon. Girders, beams, and slabs shall be placed in one operation. In walls of buildings having door and window openings, lifts of individual pours shall terminate at top and bottom of opening. Other lifts shall terminate at such levels as are indicated or as to conform to structural requirements or architectural details, or both, as directed. Special provision shall be made for jointing successive pours as indicated or required. A strip of dressed lumber shall be tacked to the inside of the forms at the construction joint. The concrete shall be poured to a point 1 inch above the under side of the strip. The strip shall be removed 1 hour after the concrete has been placed, and any irregularities in the joint line leveled off with a wood float, and all laitance removed.

## 17. Weakened-Plane Joints

Vertical control (weakened-plane) joints shown on the drawings as continuations of control joints in masonry construction, shall be sealed with an approved type of joint filler and finished off at surface with light-colored gun-type calking compound slightly groove-tooled. Floor control joints are specified under SLABS ON GRADE.

## 18. Installation of Anchorage Items

a. *Slots.* Dovetail slots shall be installed vertically in the concrete spaced not over 2 feet apart for anchoring stone or brick facing and shall be installed horizontally approximately 25 inches apart for anchoring furring. Adequate slots or inserts shall be provided for anchoring members at openings. Slots shall be provided for anchoring ends of masonry partitions abutting concrete. Where concrete columns to be faced with brick are less than 16 inches wide, slots may be omitted; where such columns are from 16 to 30 inches wide, one row of slots shall be installed; where columns are more than 30 inches wide, slots shall be installed not more than 2 feet apart on centers. Dovetail anchors may be omitted in spandrel beams less than 16 inches in depth.

b. *Wire Inserts for Plaster Accessories.* Where ribbed lath and metal furring are to be secured to the underside of concrete joists, fastenings shall be provided for each rib or furring strip at all bearings. If joists are more than 7 inches wide, two rows of fastenings shall be used so that the unsupported span will not exceed 27 inches.

c. *Inserts.* Inserts for hangers for piping and mechanical fixtures shall be as specified under section PLUMBING AND HEATING, but shall be installed under this section, in accordance with the requirements specified in sections on PLUMBING AND HEATING.

d. *Sleeper Clips.* Sleeper clips shall be spaced not more than 16 inches on centers along sleepers unless indicated.

e. *Wall Anchors.* Wall anchors for glass-block masonry shall be as specified under section MASONRY, GLASS-BLOCK, but shall be installed under this section, CONCRETE, in accordance with the requirements specified in section on MASONRY, GLASS-BLOCK.

## 19. Preparation for Placing

Water shall be removed from excavation before concrete is deposited. Any flow of water shall be diverted through proper side drains and shall be removed without washing over freshly deposited concrete. Hardened concrete, debris, and foreign materials shall be removed from interior of forms and from inner surfaces of mixing and conveying equipment. Reinforcement shall be secured in position, inspected and approved before pouring of concrete. Runways shall be provided for wheeled concrete handling equipment; such equipment shall not be wheeled over reinforcement nor shall runways be supported on reinforcement. The subgrade for hangar and warehouse floors shall be finished to the exact section of the bottom of the floor slab and shall be maintained in a smooth, compacted condition, in conformity with the required section and grade until the concrete is in place. Where concrete is placed directly on the earth fill the subgrade shall be thoroughly moistened, but not muddy, at the time the concrete is deposited.

## 20. Placing Concrete

The use of belt conveyors, chutes, or other similar equipment will not be permitted without written approval. Concrete shall be handled from mixer to transport vehicle to place of final deposit in a continuous manner, as rapidly as practicable, and without segregation or loss of ingredient until the approved unit of operation is completed. Concrete that has attained its initial set or has contained its mixing water for more than 45 minutes shall not be placed in the work. Placing will not be permitted when, in the opinion of the Contracting Officer, the sun, heat, wind, or limitations of facilities furnished by the contractor prevent proper finishing and curing of the concrete. Concrete shall be placed in the forms in uniform layers as nearly as practicable in final position. Forms or reinforcement splashed with concrete shall be cleaned in advance of pouring subsequent lifts. Immediately after placing, concrete shall be compacted by thoroughly agitating in an approved manner. Tapping or other external vibration of forms will not be permitted. Concrete shall not be placed on concrete sufficiently hard to cause formation of seams and planes of weakness within the section. Concrete shall not be allowed to drop freely more than 5 feet in unexposed work nor more than 3 feet in exposed work; where greater drops are required, a tremie or other approved means shall be employed. The discharge of the tremies shall be controlled so that the concrete may be effectively compacted into horizontal layers not more than 12 inches thick, and the spacing of the tremies shall be such that segregation does not occur. Concrete to receive other construction shall be screeded to the proper level to avoid excessive shimming or grouting.

*a. Cold-Weather Requirements.* Concrete shall not be placed when the ambient temperature is below 35° F. nor when without special protection the concrete is likely to be subjected to freezing temperature before the expiration of the specified curing period. If necessary to place concrete under conditions of low temperature, placement shall be approved by the Contracting Officer. The temperature of the concrete when placed shall be not less than 50° F. nor more than 70° F. Heating of the mixing water and/or aggregates will be required as necessary to maintain the minimum temperature of 50° F., and all methods and equipment for heating shall be subject to the approval of the Contracting Officer. Materials shall be free from ice, snow, and frozen lumps before entering the mixer. Suitable covering and other means, as approved by the Contracting Officer, shall be provided for maintaining the concrete at a temperature of at least 50° F. for not less than 72 hours after placing, and at a temperature above freezing for the remainder of the curing period. Salt, chemicals, or other foreign materials shall not be mixed with the concrete to prevent freezing, except that calcium chloride may be used as an accelerating agent only after specific approval. Any concrete damaged by freezing shall be removed and replaced at the expense of the contractor.

*b. Earth-Foundation Placement.* Concrete footings shall be placed upon undisturbed clean surfaces, free from frost, ice, mud, and water. When the foundation is on dry soil or previous material, waterproof paper, clear polyethylene sheeting 0.004 inch thick, or polyethylene-coated waterproof paper shall be laid over the surfaces to receive concrete. Above materials shall be of the type specified for curing concrete except that the polyethylene film may be clear.

*c. Rock-Foundation Placement.* Rock surfaces upon which concrete is to be placed shall be approximately level, clean, free from oil and other objectionable coatings, water, mud, debris, drummy rock, and loose semidetached or unsound fragments, and shall be sufficiently rough to assure satisfactory bond with the concrete. Faults or seams shall be cleaned to firm rock on the sides, and to a depth satisfactory to the Contracting Officer. Immediately before concrete is placed, rock surfaces shall be cleaned by high-velocity air-water jets, sand blasting, or other means satisfactory to the Contracting Officer.

*d. Chute or Conveyor Placement.* Concrete may be conveyed by chute or conveyor when permitted in writing by the Contracting Officer. The chute shall be of metal or metal-lined wood with sections set at approximately the same slope to assure a continuous uniform flow throughout the length of the chute.

The slope of the chute shall not be less than one vertical to three horizontal nor more than one vertical to two horizontal. The conveyor shall be of the bucket-lift type designed to prevent segregation of the aggregate and loss of mortar. The slope of the conveyor shall not be steeper than 45°. The discharge of the chute or conveyor shall be provided with a baffle plate or other device to prevent segregation. If the height of the discharge end is more than three times the thickness of the layer being deposited or more than three feet above the surface of the concrete in vertical forms, a spout shall be used with lower end maintained as near the surface of deposit as practicable. When pouring is intermittent, the concrete shall be discharged into a hopper. The chute and conveyor shall be thoroughly clean before and after each run. Waste material and flushing water shall be discharged outside the forms.

*e. Pump Placement.* Where concrete is conveyed and placed by pumping, the plant and equipment shall be approved by the Contracting Officer. Operation of pump shall be such that a continuous stream of concrete without air pockets is produced. When pumping is completed, concrete to be used remaining in pipeline shall be ejected without contamination of concrete or separation of ingredients. After each operation, equipment shall be thoroughly cleaned, and debris and flushing water shall be wasted outside forms.

## 21. Compaction

Concrete shall be placed in layers not over 12 inches deep, except that class P concrete shall be placed in a single layer. Each layer except lightweight concrete shall be compacted by mechanical internal-vibrating equipment supplemented by hand-spading, rodding, and tamping as directed. Vibrators shall not be used to transport concrete inside forms. Use of form vibrators will not be permitted. Internal vibrators shall maintain a speed of not less than 5,000 impulses per minute when submerged in the concrete. (At least one spare vibrator or sufficient parts for repairing vibrators shall be maintained at the site at all times.) Duration of vibration shall be limited to time necessary to produce satisfactory consolidation without causing objectionable segregation. The vibrator shall not be inserted into lower courses that have begun to set. Where absorptive form lining is used, the vibrator shall not be placed between forms and the outer row of reinforcement, and the vibrator shall not be allowed to touch the absorptive form lining.

Vibrators shall be applied at uniformly spaced points not farther apart than the visible effectiveness of the machine.

## 22. Bonding and Grouting

Before depositing new concrete on concrete that has set, the surfaces of the set concrete shall be thoroughly roughened and cleaned of laitance, foreign matter, and loose particles. Forms shall be retightened and the surfaces of the set concrete slushed with a grout coat of neat cement. New concrete shall be placed before the grout has attained initial set. The first three inches of the new concrete shall be the regular mix except that the proportion of coarse aggregate shall be reduced 50 percent.

## 23. Lightweight Structural Concrete (Class D)

Lightweight structural concrete (class D) shall be stable against deterioration or loss of strength from water, insects, rodents, or microorganisms, and shall not require excessive water or mixing.

*a. Composition.* Lightweight structural concrete shall consist of portland cement, water, and lightweight aggregate, without admixtures or aeration except air-entrainment. The weight of the dry, finished concrete shall be not more than 100 pounds per cubic foot, and the compressive strength shall be not less than 2,000 pounds per square inch at 28 days.

*b. Thermal Conductivity.* When lightweight structural concrete is used for insulation, the thermal conductivity factor shall be not more than 2.75 B.t.u. per hour per square foot per inch of thickness per degree of temperature differential, and shall be determined by the standard method of test in Federal Specification LLL-F-321. Indentation tests will not be required.

*c. Data and Samples.* Before starting work, a written statement indicating weight per cubic foot of finished concrete, exact proportions, grading of aggregate, method of mix, and installation shall be submitted for approval. Vibration, if proposed, shall be by the internal method, and the statement called for above shall include designation of type of vibration machine and recommendations of manufacturer of the lightweight aggregate regarding use of the machine. Two standard cylinders of the proposed lightweight concrete shall be furnished for tests. The test cylinders shall be made in the presence of the Contracting Officer and in accordance with ASTM Standard C 31.

*d. Reinforcement.* Reinforcing bars or mesh for lightweight structural concrete shall be as indicated and shall meet the requirements hereinbefore specified.

*e. Finish.* Finish of lightweight structural concrete shall be monolithic. For surfaces other than floors, finish shall be smooth, rubbed, or rough, as indicated and shall be produced as specified hereinafter under paragraphs on FINISHES OF CONCRETE OTHER THAN FLOOR AND ROOF SLABS. For floors and roof slabs the surface shall be screeded to a smooth even finish, wood-floated, and steel-troweled, as specified hereinafter for monolithic finishes.

*f. Curing.* Lightweight structural concrete shall be protected and cured in accordance with the requirements specified hereinafter under paragraph on CURING.

## 24. Lightweight Concrete for Fill, Other Than Insulating (Class E)

*a. Composition.* Lightweight concrete for fill other than insulating shall consist of portland cement, water, and lightweight aggregate without admixtures or aeration except air-entrainment. Weight of the dry, finished concrete shall be not more than 90 pounds per cubic foot, and the compressive strength shall be not less than 1,000 pounds per square inch at 28 days.

*b. Data and Samples.* Before work is started, a written statement indicating weight per cubic foot of finished concrete, exact proportions, and method of mixing and installing shall be submitted for approval. Vibration, if used, shall be by the internal method, and the statement called for above shall include designation of type of vibration machine, and recommendations of manufacturer of the lightweight aggregate regarding the use of the machine. Two standard cylinders of the proposed lightweight concrete, made in the presence of the Contracting Officer, shall be furnished for testing. The Contracting Officer may also take samples of concrete placed in the work for check tests.

*c. Finish.* Surface of the fill shall be screeded to the required contour and elevation and then steel-troweled to produce a smooth, dense surface skin to provide for the application of the covering specified.

## 25. Lightweight Insulating Concrete (Class F)

Lightweight insulating concrete shall be of a type that can be readily placed, finished, and cured as specified hereinafter and shall be free of any ingredients that will adversely affect adjacent materials. Expansion or ventilating joints, through the full depth of the fill, shall be installed at junction of fill with fixed vertical surfaces and at intermediate intervals not exceeding 50 feet in each direction throughout the entire roof fill. Perimeter joints shall be filled with a low-density insulating board to permit free passage of entrapped moisture vapor.

*a. Composition.* Lightweight insulating concrete shall consist of type I or type III portland cement, water, and either lightweight aggregate or chemicals or lightweight aggregate with chemicals.

*b. Proportioning and Mixing.* Lightweight insulating concrete shall be proportioned and mixed in strict accordance with the aggregate or chemical supplier's directions to produce homogeneous concrete having the following characteristics:

(1) *Thickness and density* shall be as required to provide a coefficient of heat transmission (U-value) through the completed roof construction air-to-air, not in excess of _____ B.t.u. per hour per square foot per degree F. temperature difference, when computed in accordance with recognized methods established in the latest issue of the Guide published by the American Society of Heating and Air-Conditioning Engineers. Density shall not exceed 60 pounds per cubic foot in air-dry condition.

(2) *Compressive strength* shall be not less than 125 pounds per square inch.

(3) *Indentation resistance* shall be sufficient to support a load of 140 pounds applied on a 1⅛-inch diameter disk without evidence of crushing the concrete when cured for 28 days and dried to constant weight.

*c. Expansion or Ventilating Joints.* Expansion or ventilating joints at junction with vertical surfaces shall be not less than 1 inch wide for the full depth of the fill, and shall be filled with a low-density glass fiber or mineral-wool board not exceeding 6 pound density and conforming to either Federal Specification HH-I-526 or HH-I-562. The board shall be accurately cut to conform to the contour of the fill. Intermediate joints shall be not less than ½ inch in width and shall be left open.

*d. Data.* Before the work is started, the contractor shall submit in duplicate to the Contracting Officer for approval, a written statement in accordance with the aggregate and/or chemical supplier's recommendations, stipulating the thickness and oven-dry density of the lightweight insulating concrete he proposes to provide, the exact quantities of its components, and the methods of proportioning, mixing, processing, conveying, placing, finishing, and curing required in order to produce concrete with the characteristics and requirements specified. Prior to installation the contractor shall furnish a plan and details of the edge venting, the expansion or ventilating joints, and the joint filler, for approval.

*e. Placing and Finishing.* Lightweight insulating concrete, proportioned and mixed as specified, shall be conveyed and placed in strict accordance with the aggregate or chemical supplier's directions to produce concrete of uniform density throughout and with top surface screeded or otherwise finished suitable for the reception of the roofing specified, and to a true plane within a tolerance of plus or minus $\frac{1}{4}$ inch when tested with a 10-foot-long straightedge applied in any direction. Additional surfacing material shall be applied to exposed top surfaces as may be necessary to obtain a surface uniformly suitable for the reception of the roofing specified.

*f. Curing and Protecting.* The surface of finished lightweight insulating concrete shall be adequately protected from damage from heat, cold, rain, snow, direct rays of the sun, humidity, and wind, shall be cured and allowed to dry out suitably for reception of the roofing materials specified, and shall be maintained in such condition until the roofing has been installed.

*g. Workmen and Supervision.* Lightweight insulating concrete shall be proportioned, processed, mixed, conveyed, placed, finished, cured, and protected by experienced workmen under the direct and constant supervision of a foreman or superintendent skilled in proportioning, processing, mixing, conveying, placing, finishing, curing, and protecting lightweight insulating concrete as provided under this contract.

*h. Specimens.* Test specimens shall be taken as directed for testing the insulating concrete placed during each day's operations. Specimens shall be made in the presence of the Contracting Officer who may also take additional samples for check tests. Standard 6- by 12-inch cylinders shall be furnished for determining density, for compressive-strength tests in accordance with American Society for Testing Materials Standard C 39, and for indentation tests. The specimens shall be put in molds for 48 hours with top not covered and then removed from the molds and stored 26 days in air under job conditions but protected from weather, except that if type III cement is used the specimens shall be stored for 5 days in air under job conditions following

2 days in the molds. The specimens shall be tested for compressive strength and indentation strength in the air-dry condition at the age of 28 days for type I cement and 7 days for type III cement. The end surfaces of specimens shall be prepared by sawing or grinding to a plane surface, and tested without application of a capping material. The specimens shall be ovendried to constant weight at a temperature between 220° and 230° F. for determining the ovendry density.

## 26. Slabs on Grade

*a. General.* The installation of underground and embedded items shall be approved before slabs are placed. Pipes and conduits shall be installed below the concrete unless otherwise indicated. Fill required to raise the subgrade shall be placed as specified under Section EXCAVATION, FILLING, AND BACKFILLING (FOR BUILDING CONSTRUCTION). Drainage fill not less than 6 inches in thickness shall be installed under slabs on grade. The fill shall be leveled and uniformly compacted to a reasonably true and even surface. Immediately prior to placing the concrete the fill shall be covered with a vapor barrier.

*b. Concrete.* Concrete shall have a slump of no more than 2 inches unless a greater slump is specifically authorized. Concrete shall be compacted, screeded to grade, and prepared for the specified finish. Contraction joints shall be provided in large slabs by pouring each slab in alternate checkerboard sections approximately 800 square feet in area, or at the option of the contractor the slab may be poured continuous as limited by expansion and construction joints, and contraction joints may be formed by the insertion of fiberboard strips in the wet concrete, or may be cut with an approved concrete-sawing machine after the concrete has set. Sawed joints shall be cut at a time to be determined by the Contracting Officer and shall be $\frac{1}{8}$ inch in width and approximately $\frac{1}{4}$ of the slab thickness in depth unless otherwise indicated or directed. Sludge and cutting debris shall be removed from cut joints. Fiberboard joints shall be made with a strip of $\frac{1}{8}$-inch thick hard-pressed fiberboard approximately $\frac{1}{4}$ of the slab thickness in width and as long as practicable. After the first floating, the concrete shall be grooved with a tool at the desired joint locations to a depth approximately equal to the width of the strip. The strip shall be inserted in the groove, using a U-shaped device of sheet metal fitted over the top edge of the strip to maintain alinement until the top edge of the strip is flush with the surface of the slab. When the concrete has set sufficiently to retain the strip, the sheet metal device shall be withdrawn. The slab shall be floated and finished as specified, using an edging tool on each side of the inserted joint strip. Concrete joints, where sawed or formed, shall be filled with joint-filling compound, except where floor is to receive floor covering.

*c. Vapor Barrier.* Vapor barrier shall be carefully installed to avoid puncture or tear. Punctures and tears occurring during subsequent operations shall be patched. Edges shall be lapped not less than 4 inches and end joints shall be lapped not less than 6 inches. Patches and lapped joints shall be sealed with a pressure-sensitive adhesive or pressure-sensitive tape not less than 2 inches wide. Adhesive or tape shall be compatible with the membrane, and as recommended by the manufacturer of the membrane.

## 27. Ribbed-Concrete-Slab Construction

*a. Flat-Ceiling Type.* Where indicated on the drawings ribbed concrete slabs construction with filler units to provide flat ceilings may be constructed with either hard-burned hollow-tile or hollow-concrete units 12 inches wide, 12 or 16 inches long except where shorter units are required at ends of rows, and of the depth required to form slabs of total thickness indicated. Filler units shall be free of broken corners on the under side, and broken corners on the upper side shall be covered with insect screen to prevent loss of grout. Open cells at ends of rows shall be closed by slabs of tile or concrete or by rectangular pieces of insect screening or hardware cloth not coarser than $\frac{1}{4}$-inch mesh. Blocks shall be thoroughly sprinkled with water before concrete is poured. If skeleton form work is used, the forms under the ribs shall be wide enough to provide $1\frac{3}{4}$-inch bearing on each side for the filler units. Filler units shall be laid in straight lines and closely butted. Optional types of blocks, producing units of the specified horizontal dimensions, will be considered for approval, provided that the ultimate compressive strength of the area is not less than that of the blocks specified above. If a proposed optional method is based upon the use of open-end blocks, such openings shall be closed with internal metallic closures to prevent concrete from entering the voids and at the same time permitting full contact between the poured concrete and ends of the blocks. If a proposed optional method is based upon the use of terracotta blocks or any other type of block differing appreciably in texture from concrete joists, soffit blocks shall be placed under ribs, only where plaster is to be applied directly to the slab.

*b. Removable-Metal-Pan-Form Type.* Where indicated, ribbed concrete slabs shall be constructed with metal-pan-form units true to proper cross section. Pan units shall be closely and neatly fitted together and to the splayed units adjacent to supporting beams, girders, or walls. Pan units shall be sufficiently tight to prevent loss of grout during pouring and to form a true, clean, smooth surface, free of honeycomb and rough exposed-aggregate area. Pan units for exposed ceilings in finished areas shall be of the long-span type of sufficient length to span from beam to beam or bridging joist without intermediate lapped joints. Ceilings to be left exposed in the completed structure, except in crawl spaces, unfinished attic spaces, and other rough, entirely unfinished spaces, shall be formed with well-fitting undamaged pans and shall have projecting ridges dressed off flush, honeycomb, pitting, and other minor defects pointed flush and rubbed with burlap.

## 28. Setting Column Bases

Loose and attached column base plates and bearing plates for beams and similar structural members shall be fully bedded on wedges or shims, and damp-pack bedding mortar. The bedding mortar shall be a mix composed of 1 part portland cement, $2\frac{1}{2}$ parts sand, and not more than $4\frac{1}{2}$ gallons of water per bag of cement. Sand shall have the following gradation:

Passing a No. 8 sieve____95 to 100 percent.
Passing a No. 16 sieve____65 to 90 percent.
Passing a No. 50 sieve____10 to 30 percent.
Passing a No. 100 sieve____3 to 10 percent.

The top of concrete or masonry piers or other bearing surfaces shall be finished to an elevation that is lower than the elevation of the bottom of the base or bearing plate by approximately 1/24 of the width of the base or bearing plate. The base or bearing plate shall be set and anchored to the proper line and elevation. Metal wedges, shims and/or setting nuts shall be used for leveling and plumbing the structural members, including plumbing of the columns. The space between the top of the concrete or masonry bearing surface and the bottom of the bearing plate or base shall be packed with the above cement mix by tamping or ramming with a bar or rod until voids are completely filled. Wedges or shims shall not be removed, but where protruding, shall be cut off flush with the edge of the base or bearing plate.

## 29. Finishes of Concrete Other Than Floor and Roof Slabs

Immediately after removal of the forms all fins and loose material shall be removed; honeycomb, aggregate pockets, voids, and holes over ½ inch in diameter shall be cut out to solid concrete, thoroughly wetted, brush-coated with neat cement grout, and filled with cement mortar composed of 1 part light-colored portland cement to 2 parts fine aggregate. Mortar shall be placed in layers as required, and each layer shall be thoroughly compacted in place. The final layer shall be finished flush and in the same plane as adjacent surfaces. Patchwork shall be damp cured for 72 hours. Exposed patchwork shall be rubbed or otherwise treated to match adjacent surfaces.

a. Smooth finish shall, in addition to the above, be given to all interior and exterior concrete surfaces that are either to be painted or to be exposed to view as finished work. Smooth finish shall consist in thoroughly wetting and then brush-coating the surfaces with cement grout composed of 1 part light-colored portland cement to 2 parts fine aggregate mixed with water to the consistency of thick paint. Grout shall be cork- or wood-floated to fill all pits, air bubbles, and surface holes. Excess grout shall be scraped off with a trowel and the surface rubbed with burlap to remove any visible grout film. In hot, dry weather, the grout shall be kept damp by means of fog spray during the setting period. The finish for any area shall be completed in the same day and the limits of a finished area shall be made at natural breaks in the finished surface. Painting of exposed-to-view concrete surfaces is specified under the PAINTING section of these specifications.

b. Rubbed finish shall be given to exposed-to-view surfaces throughout the exterior of the structure and to_____. Rubbed finish shall consist of smooth finish, as specified above, rubbed with carborundum stones and water. No mortar or grout shall be employed during rubbing, and mortar that is worked up during rubbing shall be removed. Form marks and similar blemishes shall be removed and the surface finish left uniformly smooth and washed clean.

c. Surfaces to receive plaster, stucco, or metallic waterproofing:

  (1) *Surface treatment.* Surfaces to which plaster, stucco or metallic waterproofing are to be applied directly shall be removed to a depth of not less than 1/16-inch by chipping with a pneumatic chisel, by retarding the setting of the surface cement with a compound and removing the surface by scouring, or by other approved method that will expose the aggregate and leave a clean, firm, rough, granular surface. Treatment shall not affect the setting or strength of the concrete beyond a depth of ⅛ inch nor prevent the setting of the surface cement within a reasonable time after the forms are removed. Soffits of concrete joists less than 6 inches wide between masonry-unit fillers need no special treatment.

  (2) *Flexible inserts.* When wood forms are used, flexible dovetail inserts may, at the option of the contractor, be used to form cavities for the mechanical bonding of the surface plaster, stucco or metallic waterproofing, in lieu of the chipping or scouring treatment of surfaces to receive plaster or stucco. The inserts shall be approximately ⅜-inch deep by 2 inches in diameter, attached to the forms 12 inches on centers in accordance with the manufacturer's directions, and shall be easily removable without damaging the concrete.

d. The contractor shall pour for approval a sufficient number of sample concrete panels to show the surface finishes required. Each panel shall be not less than 6 feet long by 4 feet high. Pouring of concrete requiring the finish indicated by the samples shall not proceed until the sample panel has been approved.

## 30. Concrete Floor and Roof-Slab Finishes

Finished floor- and roof-slab surfaces shall be true plane surfaces, with a tolerance of ⅛-inch in 10 feet unless otherwise indicated. Surfaces shall be pitched to drains. The dusting of finish surfaces with dry materials will not be permitted. Concrete and mortar setting beds and fills required in connection with ceramic floor tile are covered under Section TILE WORK, CERAMIC; FOR FLOORS AND WALLS of these specifications.

a. *Monolithic Finish.* Except where otherwise specified, floor and roof slabs shall be finished by tamping the concrete with special tools to force the coarse aggregate away from the surface, then screeding and floating with straightedges to bring the surface to the required finish level. While the concrete is still green but sufficiently hardened to bear a man's weight without deep imprint, it shall be wood-floated to a true, even plane with no coarse aggregate visible. Sufficient pressure shall be used on the wood floats to bring moisture to the surface. After the surface moisture has disappeared, surfaces shall be steel-troweled to a smooth, even, impervious finish, free from trowel marks. After the cement has set enough to ring the trowel, the surface of all slabs shall be given a second steel-troweling to a burnished finish except roof slabs, hangar- and warehouse-floor slabs, and slabs to receive resilient or wood-block flooring or membrane waterproofing. *Contractor's option:* In order to facilitate finishing procedures specified, the contractor may, at his option, screed the slab surface as specified to a true even plane ½ inch below the required finish level, and, while the concrete is still green, install a ½-inch-thick surfacing mixed in proportions, by volume, of 1 part portland cement, 1 part fine aggregate, and 2 parts coarse aggregate. The fine aggregate shall pass a ¼-inch-mesh sieve, with not more than 5 percent passing a 100-mesh sieve, and not more than 10 percent passing a 50-mesh sieve. The coarse aggregate shall be graded from ⅛-inch to ⅜-inch in size, with at least 95 percent passing a ⅜-inch-mesh sieve and not over 10 percent passing a No. 8 sieve. The water shall not exceed 5 gallons per bag of cement. The surfacing shall be installed to a uniform thickness of ½ inch and finished in all respects the same as specified for monolithic finish.

*b. Rough Slab Finish.* Slabs to receive fill and mortar setting beds shall be finished by tamping the concrete with special tools to force the coarse aggregate away from the surface, and screeding with straightedges to bring the surface to the required finish plane.

*c. Broomed Finish.* Exterior concrete slabs, ramps and load platforms, and interior traffic aisles subject to motorized traffic shall be finished by tamping the concrete with special tools to force the coarse aggregate away from the surface, screeding and floating to bring the surface to the required finish level, steel-troweling to an even smooth surface, and brooming with a fiber-bristle brush in a direction transverse to that of the main traffic.

*d. Heavy-Duty Finish.* Floors or slabs to receive heavy-duty finish shall be screeded to a true, even plane ¾-inch below the required finish level, and while the concrete is still green a hard-aggregate wearing surface shall be applied.

    (1) *Mix.* Concrete for heavy-duty surfaces shall consist of selected aggregates, portland cement, water, and additives where so indicated in the aggregate producer's basic formulation. Proportioning shall be in strict accordance with the aggregate pro-

ducer's printed data. Coarse aggregate shall be uniformly graded diabase basalt, emery, granite, or other natural or manufactured aggregate having equivalent hardness and wearing qualities, and ½-inch maximum size. Maximum water content shall be 3¾ gallons per bag of cement.

    (2) *Placing and finishing.* The ambient temperature of the space to receive the heavy-duty finish shall not be not less than 50° F. Immediately after placing, the wearing surface coat shall be screeded to a true, even plane and thoroughly tamped by mechanical equipment or rolled with rollers weighing not less than 10 pounds per linear inch of roller width to obtain maximum density for full depth of the wearing-surface coat. After tamping or rolling, abrasive aggregate shall be sprinkled uniformly over the surface at the rate of ¼ pound per square foot. After sprinkling the surface with aggregate there shall be not less than two steel trowelings. The first troweling shall remove all marks and blemishes and bring the surface to a hard impervious finish. Subsequent troweling shall produce a ringing sound from the trowel and leave a burnished finish.

    (3) *Curing and protection.* Immediately following the final troweling, the finished surface shall be dampened and covered with concrete curing paper, mats, or burlap. Edges and end laps of paper shall be sealed and mats, or burlap shall be kept damp for not less than 10 days. Use of the floor shall not be permitted for at least 5 days after completion of curing and only light use shall be permitted for an additional 10-day period.

*e. Nonslip Finish.* Concrete stair treads, platforms, ramps, and loading docks in all buildings and main entrance aisles of warehouses shall have a nonslip finish. Finish of exterior surfaces shall be obtained by brooming with a stiff-fibered broom. Finish of interior surfaces shall be obtained by evenly sprinkling not less than ¼ pound of abrasive material over each square foot of the finished surface after screeding and floating, just prior to troweling. Abrasive aggregate shall be steel-troweled to a smooth, even finish. Interior stair treads and platforms shall be given a second steel troweling. Where the abrasive is covered with laitance or cement coating during troweling, the surface shall be brushed after 7 days with a steel brush, rubbed with an abrasive stone, or sand-blasted to expose the abrasive particles.

*f. Roller-And-Screed Finish (Option).* In lieu of screeding and floating as specified, the contractor may use roller-and-screed method in finishing floor slabs to receive resilient and wood-block floorings, ceramic tile set in adhesive, membrane waterproofing, or other applied flooring, whereby the slab shall be finished by tamping with special tools to force the aggregate away from the surface and then brought to the required height by a roller not less than 6 inches in diameter and of an approved type. The roller shall be carried on screeds accurately set in the work and sufficiently rigid to withstand the rolling. Concrete shall be maintained ahead of the roller at a height to insure complete compaction of the surface. Areas not reached by rolling shall be thoroughly tamped. After surface water has disappeared and the concrete is sufficiently firm to hold a man's weight, the floor shall be steel-troweled to a true, even, smooth surface.

*g. Power-Machine Finish (Option).* In lieu of hand finishing, the contractor may use an approved power finishing machine in accordance with the directions of the machine manufacturer. The preparation of concrete surfaces for finishing by machine shall in general be as hereinbefore required for handfinishing.

*h. Vacuum Processing (Option).* The vacuum process or its equivalent may be used at the contractor's option in dehydrating and compacting concrete slabs and in applying floor finishes. The vacuum process or its equivalent shall dehydrate the concrete and simultaneously compact it with a pressure of not less than 1,500 pounds per square foot to that the concrete will be of no-slump consistency immediately after processing. Control shall be adequate to limit variations in slump to not more than 1 inch in batches of the same pour. Before placing the concrete, screed strips shall be set not over 10 feet apart. The concrete shall be screeded, and immediately thereafter the vacuum process or its equivalent shall be applied for 1 or 2 minutes per inch of slab thickness. No other finishing operations will be required where slabs are to receive fill or other floor finishes. After processing, slabs that are to receive concrete finishes and rough slabs that are to receive waterproofing or roofing shall be power-floated to a true, uniform surface. The floor surface shall then be checked with a straightedge to assure that a tolerance of plus or

minus ⅛ inch in 10 feet is maintained. Troweling operations shall then be performed for the various floor finishes hereinbefore specified.

## 31. Concrete Base

Concrete base shall be installed in _____. Base shall be composed of 1 part portland cement to 2 parts fine aggregate, troweled to a smooth even surface. Reinforcing shall be of 18-gage wire, 2 x 2 meshes to the inch unless otherwise shown. Base shall be coved to the floor only where so indicated.

## 32. Curing

Curing shall be accomplished by preventing loss of moisture, rapid temperature change, and mechanical injury or injury from rain or flowing water for a period of 7 days when normal portland cement has been used or 3 days when high-early-strength portland cement has been used. Curing shall be started as soon as free water has disappeared from the surface of the concrete after placing and finishing. Curing of formed undersurfaces of beams, girders, floor slabs, and other similar undersurfaces, shall be accomplished by moist curing with forms in place for the full curing period, or, if forms are removed prior to the end of the curing period, by other approved means. Curing may be accomplished by any of the following methods or combination thereof, as approved.

*a. Moist Curing.* Unformed surfaces shall be covered with burlap, cotton, or other approved fabric mats, or with sand, and shall be kept continually wet. Where formed surfaces are cured in the forms, the forms shall be kept continually wet. If the forms are removed before the end of the curing period, curing shall be continued as on unformed surfaces, using suitable materials. Burlap shall be used only on surfaces which will be unexposed in the finished work and shall be in two layers.

*b. Waterproof-Paper Curing.* Surfaces shall be covered with waterproof paper lapped 4 inches at edges and ends, and sealed with mastic or pressure-sensitive tape not less than 1½ inches wide. Paper shall be weighted to prevent displacement, and tears or holes appearing during the curing period shall be immediately repaired by patching.

*c. Membrane-Forming Curing.* Membrane curing compound shall be applied by power spraying equipment using a spray nozzle equipped with a wind guard. The compound shall be applied in a two-coat, continuous operation at a coverage of not more than 200 square feet per gallon for each coat. When application is made by hand sprayers, the second coat shall be applied in a direction approximately at right angles to the direction of the first coat. The compound shall form a uniform, continuous, adherent film that shall not check, crack, or peel, and shall be free from pinholes or other imperfections. Surfaces subjected to heavy rainfall within 3 hours after compound has been applied, or surfaces damaged by subsequent construction operations within the curing period shall be resprayed at the rate specified above. Membrane curing compound shall not be used on surfaces that are to receive bituminous membrane waterproofing, resilient floor covering, ceramic tile set in adhesive, concrete fill or setting beds, nor on surfaces that are to be painted. Surfaces coated with curing compound shall be kept free of foot and vehicular traffic and other sources of abrasion during the curing period.

*d. Polyethylene Sheeting and Polyethylene-Coated Waterproof Paper.* Surfaces shall be completely covered. Where single sheet does not cover entire surface the ends and edges shall be lapped not less than 4 inches and sealed with adhesive tape conforming to Federal Specification PPP–T–60, type III, class 1.

## 33. Platform Fills

Fills under lockers, raised closet floors, and elsewhere as shown on the drawings shall be formed to proper heights and areas. The rough fill shall be lightweight concrete (class E) laid straight and level. Top surfaces shall be finished with not less than a $\frac{1}{2}$-inch smooth integral cement finish as indicated. Bases shall be cement as specified herein for cement base, or the surfaces shall be prepared for other base material to match the rooms in which they occur or which they adjoin.

## 34. Fill Between Sleepers

*a. Slabs on Grade and Gymnasium Floors.* Fill between sleepers for wood-floor construction over slabs on earth or on drainage fill on earth and fill between sleepers and gymnasium floors shall be tar concrete or asphalt concrete at the option of the contractor. The fill shall be composed of not less than 40 nor more than 50 gallons of tar or asphalt to each cubic yard of sand. Before mixing, the sand shall be thoroughly tried by heating, and the tar or asphalt shall be heated to such temperature that when mixing takes place, the temperature will not exceed 225° F. A $\frac{1}{2}$-inch-thick layer of the mixture shall be placed on the slab, and after the sleepers specified under the section on CARPEN-

TRY have been set, leveled, and bedded, the space between the sleepers shall be filled with the mixture. The fill shall then be tamped to a level finish $\frac{1}{2}$ inch below the top of the sleepers.

*b. Slabs Above Grade.* Where concrete fill between sleepers for woodfloor construction except gymnasium floors, is to be used on reinforced concrete slabs above grade, the fill shall be composed of one part portland cement, $2\frac{1}{2}$ parts sand, and not over 4 parts pea gravel or crushed stone passing a $\frac{1}{2}$-inch screen; or may be composed of 1 part portland cement to not over $4\frac{1}{2}$ parts of fine gravel and sand mixture. The water-cement ratio shall be the minimum necessary to produce a workable concrete mix. After the sleepers have been properly set in the clips as specified under Section CARPENTRY, the sleepers shall be solidly bedded by depositing the concrete and spading it under the sleepers. Spaces between sleepers shall then be filled to within $\frac{1}{2}$-inch of top of sleepers with the concrete fill, and the concrete shall be screeded or roughly floated to a uniform surface.

## 35. Fill Over Wood Subflooring

Fill over wood subflooring shall be class C concrete, reinforced with 4- by 4-inch mesh steel-wire fabric of not lighter than 11-gage wire unless otherwise indicated. The fill shall be placed over a layer of waterproof sheathing paper, provided as specified under Section CARPENTRY.

## 36. Setting Beds

Setting beds required over slabs for floor finishes other than concrete are covered under other sections of these specifications in which the floor finish is specified, and do not form a part of the work under this section.

## 37. Precast Concrete Trim

Precast concrete trim items are specified under section MASONRY.

## 38. Ornamental Concrete

Ornamental concrete shall be precast and of the same type concrete as adjacent areas. Where required by the ornamental detail, the maximum size of the coarse aggregate shall be reduced. Precast ornament shall be set into the structural concrete not less than 2 inches, and shall be placed during pouring of the structural concrete. For other than simple molded work, a model of wood, plaster, or other suitable substance shall be submitted for approval. Waste molds of plaster or wood shall be made from the approved model. Plaster molds shall be reinforced with jute fiber and necessary bracing. Wood molds shall be of clear lumber kerfed in back for easy removal, shall be wedged between projections, and shall have joints arranged to be inconspicuous in the finished surface. Molds shall be constructed of simply shaped and easily handled sections. Surfaces of molds designed for concrete contact shall be shellacked with two shop coats and one coat after assembly of mold.

## 39. Pipe-Trench Covers

a. *Removable covers* shall be cast in metal frames as indicated, and shall be finished the same as the adjacent floor surfaces. Unless otherwise indicated, trench covers shall be cast in approximately 4-foot-long sections and shall be provided with lifting arrangements as indicated or as approved.

b. *Fixed covers* shall be either precast or cast-in-place to the thickness indicated and in approximately 4-foot lengths. Covers shall fit into a rabbet in the trench wall and shall be separated therefrom by a double thickness of waterproof sheathing paper on the bottom and by a mastic strip or other approved material along the edges. Cement floor finish or other floor covering shall extend over the trench covers to provide a continuous floor finish.

## 40. Payment

No separate payment will be made for the work covered under this section, and all costs in connection therewith shall be included in the lump-sum contract price for the structure to which the work pertains. There will be no additional compensation for changing the proportions of the mix to overcome field deficiencies and to obtain the specified qualities and characteristics of the concrete.

# NOTES TO CONTRACTING OFFICER

1. This specification is to be used in the preparation of contract specifications. It will not be made a part of a contract merely by reference; pertinent portions will be copied verbatum into the contract documents.
2. The section number will be inserted in the specification heading and will be prefixed to the paragraph and page numbers.
3. Paragraph 2: See note 7 below.
4. Paragraph 3: The listed designations for publications are those that were in effect when the specification was being prepared. These designations will be changed, as required, to those in effect on the date of invitation for bids, and the nomenclature, types, grades, classes, and so on, referenced in the specifications should be checked for conformance to the latest revision or amendment. To minimize the possibility of error, the letter suffixes, amendments, and dates indicating specific issues will be omitted elsewhere in the specification. It is essential, therefore, that the list of applicable publications be retained in the contract specifications.
5. This specification covers the concrete work for building structures only. It is not intended for use for wharves, docks, dams, bridges, roads, walks, runways, and other engineering works including large structures of mass concrete.
6. Specification requirements for class P concrete are included in this specification, and are applicable to hangar floors, aprons, and hardstandings included in building construction, and to warehouse and heavy shop floors on grade subject to abnormally heavy loading and traffic. Such floor slabs will be designed on the basis of the flexural strength of 90-day-old concrete, and the thickness required to meet local conditions will be indicated on the drawings.
7. The following paragraphs require insertions applicable to the particular structure proposed for contract:

    2: Include herein concrete work shown on drawings but placed by other trades. The paragraph may be deleted where the possibility of conflict as to the responsibility for placement does not exist.

    5b: Insert the maximum size of coarse aggregate, based on the design, spacing, and clearances of reinforcement. The coarse aggregate for floor slabs on grade will be of maximum size not greater than $\frac{1}{3}$ the thickness of the slab, but in no event will the nominal size aggregate exceed $2\frac{1}{2}$-inches. Delete inapplicable columns from tabulation for gradation of aggregates for class P concrete.

    11a: If class P concrete is to be used, insert the 28-day flexural strength required, based on the relation between the 28-day and 90-day strengths for the aggregate proposed for use.

    11b: Insert the locations of various types of concrete. Delete types not required.

    14a: Insert the capacity of mixer in cubic yards in 8 hours.

    14c: Insert capacity of mixer in cubic feet.

    25b(1): Insert maximum allowable coefficient of heat transmission (note 27).

    29b: Insert surfaces to receive rubbed finish.

    30b, c, d, and e: List the spaces to receive the various floor finishes.

    31: Insert spaces to have concrete base.

8. Paragraph 5g(1): For projects with alkali-reactive aggregates add the following: "The alkali content shall not exceed 0.6 percent."
9. Paragraph 5i(4): Testing of very small quantities of curing compound is not considered justified. Areas up to 1,000 square feet (requiring the use of approximately 5 gallons of compound) will be limited to the other specified methods of curing. Paragraph 32c will be changed or omitted accordingly.
10. Paragraph 5j and the requirements for drainage fill in paragraph 26 may be deleted where subsurface material is of such granular formation as to preclude capillarity, and for fills under floors and slabs at loading elevations. Where drainage fill is deleted, a vapor barrier will be placed under the slab, and paragraph 26a modified accordingly.
11. Paragraph 5r(1): If the columns have been designed for hard steel, the project specification will call only for grade E or G steel for the vertical bars in columns.

12. Paragraph 5r(2): The project specifications will call for the type of spiral steel for which the project was designed.

13. Paragraph 6a: The use of an air-entraining agent will be mandatory only where an impervious weather-resistant concrete is desired. In other portions of the work, its use will be optional with the contractor.

14. Paragraph 6b: Where an accelerating agent is authorized, the amount of calcium chloride will normally not exceed 2 percent, by weight, of the portland cement, whether dry form or standard solution is used. One percent is sufficient in most cases. Calcium chloride should not be used where stray electric currents are expected, or in prestressed concrete.

15. Paragraph 7b: Where cement is to be used in small quantities, (1,200 bags or less), and structural design strength is not critical, cement may be accepted on the basis of the manufacturer's certificate of compliance with the specification, in which case the portion of the paragraph pertaining to large quantities of cement will be omitted. For projects requiring more than 1,200 bags, the portion of the paragraph requiring tests will be used and the portion pertaining to small quantities of cement will be omitted. The Contracting Officer may in cases of urgency waive the 28-day tests and permit the use of cement that has satisfactorily passed the chemical soundness, and 7-day strength tests. *The boxed notes are for identification only and will be omitted from project specifications.*

16. Paragraph 7e: A job curve to determine maximum allowable water content is illustrated in the Portland Cement Association publication, "Design and Control of Concrete Mixtures," ninth edition, 1948, page 8.

17. Paragraph 7e(1): This requirement may be modified at the discretion of the Contracting Officer to require only one or two sets of three test specimens, depending on the volume of concrete placed. If class P concrete is not included in this section, parenthetical portions for the testing of beams for flexural strength will be deleted.

18. Paragraph 10a(7): The requirement for encasing steel columns and main girders is intended to apply only where concrete fireproofing is to be used and not preclude the use of other fireproofing materials.

19. Paragraph 10b: Where climatic conditions will result in the staining of concrete due to rusting of metal inserts and reinforcing steel supports, chairs, and spacers, such accessories will be galvanized or otherwise treated.

20. Paragraph 11a and 11b: See note 7.

21. Paragraph 12c: This paragraph places the responsibility for the design of concrete mixes upon the contractor. On major building projects, including hangar and large warehouse floors, where it is believed that better results can be obtained by placing the responsibility for the design of concrete mixes with the Contracting Officer, paragraphs 12c and 40 will be changed as follows:

"12c. *Control.* The design of the concrete mixture, to meet strength requirements of the class or classes of concrete specified, shall be the responsibility of the Contracting Officer. In designing the mix, aggregate proposed for use by the contractor and approved by the Contracting Officer will be used. The design mix will be determined well in advance of commencement of the work so as to cause no delay.

"(1) *Mix design.* Before placing any concrete, adequate quantities of the concrete ingredients proposed for use shall be supplied to the Contracting Officer for making trial design mixes. In case of change in source or character of concrete ingredients after concrete placing has started, sufficient quantities of ingredients, including the new material, shall be furnished the Contracting Officer for determining a new mix. No substitution shall be made in the materials used in the work without approval of the Contracting Officer. Average cement content will be approximately as follows:

| Class of concrete | Average cement content, bags per cubic yard |
|---|---|
| AA (3750 psi compressive strength) | 6.0 |
| A (3000 psi compressive strength) | 5.5 |
| B (2500 psi compressive strength) | 5.0 |
| C (2000 psi compressive strength) | 4.5 |
| P (___ psi flexural strength)[1] | 6.0 |

[1] Insert strength value used in paragraph 11a.

"(2) *Slump test.* Consistency will be determined by the slump test, in accordance with CRD–C 5. The slump shall fall within the following limits:

| Type of structure | Slump for vibrated concrete, inches | |
| --- | --- | --- |
| | Minimum | Maximum |
| General building construction² | 2 | 3 |
| Thin reinforced walls² | 3 | 4 |
| Heavy-duty floor and slab construction (class P concrete) | 1 | 2 |

² When pacing of concrete without vibration is approved, slump shall be from 3 to 6 inches.

"(3) *Mix proportions.* Preliminary mix proportions will be furnished the contractor by the Contracting Officer before start of operations. Adjustments will be made by the Contracting Officer, as required, to determine final proportions that will best satisfy job requirements and use of materials. Subsequent adjustment in these final mix proportions will be made by the Contracting Officer as required to compensate for variations in the gradation and moisture content of the aggregates. Necessary revisions in water-cement ratio and concrete-mix proportions shall be made as directed.

"(4) *Workability.* The consistency of the mixture will be that required for the specific conditions and methods of placement. The slump shall not exceed that specified above.

"(5) *Strength tests.* The Contracting Officer will determine the strength of the concrete in the completed work during the progress of construction by test specimens made, cured, and tested as specified herein under paragraph, SAMPLES AND TESTING. Modifications of the design mix, if required, will be made by the Contracting Officer on the basis of the strength of these test specimens.

"40 PAYMENT. Except as hereinafter provided, no separate payment will be made for work covered under this section, and all costs in connection therewith shall be included in the lump-sum contract price for the structure to which the work pertains. There will be no additional compensation for changing the proportions of the mix to overcome field and aggregate deficiencies or to obtain the specified qualities and characteristics of the concrete. Adjustment in payment will be made at contract unit-price bid for cement more or less than the average bags per cubic yard specified hereinbefore for a given mix."

22. Paragraph 12c(2): Slump limitations for class P concrete, vibrated or nonvibrated, may be adjusted at the discretion of the Contracting Officer to accommodate project conditions.

23. Paragraph 14: Where the size and type of the project does not warrant the use of a batching plant and/or mixing equipment, except truck mixers, this paragraph may be modified to require such plant and equipment as the project demands. Where the paragraph is retained, fill in blank in paragraph 14c, indicating the minimum size of mixer acceptable.

24. Paragraph 16: In localities where extreme conditions of heat and/or dryness are liable to produce excessive shrinkage, the unit of operation may be limited in size to 40 feet in any horizontal direction.

25. Paragraph 16: The drawings will include locations of all horizontal and vertical construction joints where such joints are critical with respect to design considerations.

26. Paragraph 20: Concrete placed during hot weather should have the lowest temperature which it is practicable to produce under the current conditions. If desired, the pouring of exposed slabs during the daylight hours of summer may be limited by specifying the maximum atmospheric temperature or combination of temperature, humidity, and wind velocity under which such concrete may be placed. If desired, placement of floor slabs may be delayed until protected by roof construction.

27. Paragraph 21: The requirement in parentheses for a spare vibrator may be deleted when impracticable for small jobs.

28. Paragraph 25b(1): The maximum allowable coefficient of heat transmission through the completed roof air-to-air, or U-value, will agree with that specified in the BUILD-UP ROOFING Section of the specifications, for

the board-type insulation for which the light-weight insulating concrete is an option.

29. Paragraph 25c: Where lightweight insulating fill is installed, expansion or ventilating joints are mandatory to overcome expansion in perlite and cellular concrete fills and to permit dissipation of moisture in vermiculite fills.

30. Paragraph 25d: The thickness and oven-dry density of the light-weight-insulating-concrete fill proposed by the contractor will comply within reasonable limits with those approximated by the following formula, reflecting the relation between density and thermal conductivity of insulating concretes on the basis of figure 20 in the American Concrete Institute's paper (Titles Nos. 50-48a and 50-48b) "Cellular Concretes." "D" designates density of the hardened concrete in the oven-dry condition expressed by weight in pounds per cubic foot and "k" designates its thermal conductivity expressed in B.t.u. per hour per inch thickness per square foot per degree F. temperature difference.

$$D = 35 + 89 \log_{10} k$$

31. Paragraph 25e: Nailing strips for installation of roofing or sheet metal will not be required in connection with lightweight insulating concrete.

32. Paragraphs 26 and 26a: See note 10.

33. Paragraph 26b: In hot, dry climates, or elsewhere as may be determined and advisable, the area of unit pours in slabs on grade may be limited to 600 square feet.

34. Paragraph 28: In order to improve the quality of finish on ribbed concrete construction of this type that is to be left exposed, long-span-type metal pans are specified. The requirements of this paragraph may be relaxed to permit the use of short pan forms if it is determined that long-span forms are not available or economically feasible. Forming of work that will be concealed will be by long forms or short forms at the option of the contractor. Shop drawings will be requested for approval.

35. Paragraph 29b: Rubbed finish will be employed only in monumental-type construction where a very high quality finish is essential.

36. Paragraph 30d: When a heavy-duty cement finish is specified, the contractor will be required to furnish evidence that he has installed such finishes with satisfactory results over a period of years. This requirement will be inserted in the bid form.

37. Paragraph 30h: Vacuum processing of concrete will be permitted as an option in project specifications where the structure is designed with concrete slab construction of uniform thickness, where large areas of cement finishing over concrete slabs are required, or where heavy-duty cement finish is required.

38. Paragraph 32c: See note 9.

39. Where dense floors are required following paragraph for floor-hardener application will be inserted directly after paragraph 39:

"HARDENER APPLICATION. Hardener shall be applied to concrete floors of the following spaces:

The floors shall be thoroughly cured, cleaned, and perfectly dry with all work above them completed. Zinc and/or magnesium fluosilicate shall be applied evenly, using three coats and allowing 24 hours between coats. The first coat shall be $\frac{1}{3}$ strength, the second coat $\frac{1}{2}$ strength, and the third coat $\frac{2}{3}$ strength. Each coat shall be allowed to remain wet on the concrete surface for 15 minutes. Sodium silicate shall be applied evenly, using three coats, allowing 24 hours between coats. The sodium silicate shall be applied full-strength at the rate of $\frac{1}{3}$ gallon per 100 square feet. Approved proprietary hardeners shall be applied in conformance with the manufacturer's instructions. After the final coat is completed and dry, any surplus hardener shall be removed from the surface by scrubbing and mopping with water."

40. The first sentence of the PAYMENT paragraph will be deleted from any specification contemplating one lump-sum contract price for the entire work to be covered by the invitation for bids.

# APPENDIX III

## SPECIFICATION FOR MASONRY

### 1. Scope

This section covers masonry, complete.

### 2. Applicable Publications

The following publications of the issues listed below, but referred to thereafter by basic designation only, form a part of this specification to the extent indicated by the references thereto:

*a. Federal Specifications.*

| | |
|---|---|
| HH–B–671d | Brick; Refractory, Fire-Clay |
| QQ–S–632 | Steel Bar, Reinforcing, (for) Concrete. |
| QQ–W–461b & Am–1 | Wire, Steel (carbon), Bare and Coated. |
| SS–T–341a | Tile; Structural, Clay, Load-bearing Wall. |

*b. American Society for Testing and Materials Standards.*

| | |
|---|---|
| A 153 | Zinc Coating (Hot-Dip) on Iron and Steel Hardware. |
| C 33 | Concrete Aggregates. |
| C 34 | Structural Clay Load-Bearing Wall Tile. |
| C 55 | Concrete Building Brick. |
| C 56 | Structural Clay Non-Load-Bearing Tile. |
| C 62 | Building Brick (Solid Masonry Units Made From Clay or Shale). |
| C 67 | Sampling and Testing Brick. |
| C 73 | Sand-Lime Building Brick. |
| C 90 | Hollow Load-Bearing Concrete Masonry Units. |
| C 112 | Sampling and Testing Structural Clay Tile. |
| C 126 | Ceramic Glazed Structural Clay Facing Tile, Facing Brick, and Solid Masonry Units. |
| C 129 | Hollow Non-Load-Bearing Concrete Masonry Units. |
| C 145 | Solid Load-Bearing Concrete Masonry Units. |
| C 216 | Facing Brick (Solid Masonry Units Made From Clay or Shale). |
| C 266 | Time of Setting of Hydraulic Cement by Gillmore Needles. |
| C 270 | Mortar for Unit Masonry. |
| C 331 | Lightweight Aggregates for Concrete Masonry Units. |
| C 426 | Drying Shrinkage of Concrete Block. |
| C 427 | Test for Moisture Condition of Hardened Concrete by the Relative Humidity Method. |
| E 119 | Fire Tests of Building Construction and Materials. |

*c. Underwriters' Laboratories, Inc., Publication.*
Fire Protection Equipment List (Current Issue).

### 3. Materials

*a. Aggregate.* Aggregate used in making concrete masonry units, concrete brick, and split block shall conform to ASTM Standard C 33 or C 331, except as modified hereinafter. Grading of aggregates as stipulated in Section 7 in ASTM Standard C 33 and testing of lightweight aggregates for drying shrinkage as stipulated in Section 6(a) in ASTM Standard C 331 will not be required. Lightweight aggregates shall comply with the following requirements when tested for stain-producing iron compounds:

(1) When determined by visual classification method, the iron stain deposited on the filter paper shall not exceed the "light-stain" classification.

(2) When determined by chemical-analysis method and reported as $Fe_2O_3$, the iron stain deposited on the filter paper from a 200-gram sample shall not exceed 1.2 mg. $Fe_2O_3$.

*b. Anchors and Ties.* Anchors and ties shall be of approved design and, except as otherwise specified herein, shall be zinc-coated ferrous metal of the types noted below. Zinc coating of anchors and ties shall conform to ASTM Standard A 153, Class B–1, B–2, or B–3, as required.

(1) *Wire mesh ties* for anchorage of 4-inch thick partitions to exterior walls shall be made of steel wire not lighter than 0.0625 inch (16 gage) in diameter and shall be 3 inches wide, of an effective length for this purpose, and with $\frac{1}{2}$-inch mesh.

(2) *Wire ties* for anchoring concrete-masonry-unit partitions to exterior walls shall be of joint-reinforcement segments with longitudinal wires bent as necessary to provide bond equivalent to a cross wire at each end.

(3) *Corrugated or crimped metal ties* shall be not less than $\frac{7}{8}$ inch wide and of sheet steel not lighter than 0.0299 inch in nominal thickness (22 gage).

(4) *Ties for cavity-type walls* shall be rectangular in shape, of the required length not less than 4 inches wide, and made of nominal 3/16-inch-diameter wire. The wire shall be commercial bronze, corrosion-resisting steel or copper-clad steel, the copper constituting at least 25 percent of the cross-sectional area of the copper-clad steel wire, and shall be crimped in the center to provide an effective moisture drip.

(5) *Dovetail-type anchors for use with embedded slots or inserts* shall be of sheet steel not lighter than 0.0598 inch (16 gage) in thickness by 1-inch-wide flat anchors for concrete masonry units and structural-clay-tile facing, and steel wire not lighter than 0.1483 inch (9 gage) in diameter for brick and split-block facing. Dovetail slots and inserts are specified under SECTION: CONCRETE.

(6) *Rigid steel anchors* for anchoring interior bearing walls, partitions over 4 inches thick and fire walls to exterior walls shall be 1 inch wide by 1/4 inch thick with each end turned down not less than 3 inches for setting into filled cells, and with not less than 24 inches between the downturned ends.

*c. Joint Reinforcement.* Joint reinforcement shall be made of hard-temper, zinc-coated steel wire conforming to Federal Specification QQ-W-461, type I, finish 4, zinc-coated, the wire to be zinc-coated before being used in reinforcement. Longitudinal wires may be smooth or deformed and shall be not lighter than 0.1620 inch (8 gage) in diameter. Cross wires shall be not lighter than 0.1055 inch (12 gage) in diameter. The distance between contact of cross wires with longitudinal wires measured on each longitudinal wire shall not exceed 6 inches for smooth longitudinal wires, and 16 inches for deformed longitudinal wires. The spacing of the longitudinal wires in joint reinforcement used with load-bearing and non-load-bearing units shall be 2 inches less in the load-bearing units and 1½ inches less in the non-load-bearing units, than the nominal width of the respective block. Cross wires may be placed between and in the same plane as deformed longitudinal wires, but shall intersect above or below plain longitudinal wires with ends of cross wires extending not more than 1/8 inch beyond the outer sides of the longitudinal wires. Joint reinforcement shall be furnished in flat sections ranging from 10 to 20 or more feet in length. Reinforcement furnished in rolls will not be permitted.

*d. Brick.* Brick shall be common clay or shale brick conforming to ASTM Standard C 62, except that brick for exterior facing shall be selected for uniformity in size and color. At the option of the contractor, brick for exterior facing may conform to ASTM Standard C 216, type FBS, selected for color range of approved sample. Grade SW shall be used for exterior facing work below grade and for the first 6 courses above grade; and either grade SW or MW at the contractor's option shall be used for all other work.

*e. Concrete Brick and Split Block.* Concrete brick and split block conforming to ASTM Standard C 55, except as hereinafter modified, may be used, at the option of the contractor, in lieu of common clay or shale brick, except that concrete brick and split block shall not be used for exterior facing unless such facing is indicated as concrete brick or split block. Units shall be either grade A-1, A-2, B-1, or B-2 as specified in table I below, except that only grade A-1 units shall be used for work below grade. Concrete brick and split-block units that have been subjected during manufacture to a saturated steam pressure of 120 pounds or more per square inch for 5 hours or more may be used as soon as cooled. Units that have not been subjected to such steam pressure shall be cured for 28 days or more before delivery. Units shall be delivered to the job site in an air-dry condition, shall conform to grade and physical requirements, and shall be classified into group 1 or group 2 depending upon the linear-shrinkage potential as stipulated in table 1 below.

*Table I. Linear Shrinkage Potential*

| Grade | Application | Minimum compressive strength (brick, flatwise) psi average gross area[1] | Concrete density lbs. per cu. ft. | Maximum absorption lbs. per cu. ft.[2] | Maximum linear shrinkage, percent (ASTM Standard C 426) | |
|---|---|---|---|---|---|---|
| | | | | | Group 1 | Group 2 |
| A–1 | Exterior grade (unprotected). Face or veneer brick not to be painted or otherwise protected from weather. | 3,500 | 120 or more | 9 | 0.065 | 0.025 |
| A–2 | Exterior grade (protected). Face or veneer brick to be painted or otherwise protected from weather. | 2,500 | 120 or more | | .065 | .03 |
| | | 2,500 | Less than 120 | | .065 | .04 |
| B–1 | Backup grade (load-bearing) | 2,500 | 120 or more | | .065 | .04 |
| | | 1,500 | Less than 120 | | .065 | .04 |
| B–2 | Backup grade (non-load-bearing) | 1,500 | All densities | | .065 | .04 |

[1] Average of 5 bricks. The strength of individual brick shall be not less than 80 percent of the average strength of 5 bricks.
[2] Average of 5 bricks.

*f. Concrete Masonry Units.* Concrete masonry units shall be of modular dimensions and shall include all closers, jamb units, headers, and special shapes and sizes required to complete the work as indicated. Throughout habitable spaces vertical external corners that are exposed-to-view or painted shall be bullnose. Units exposed to view or painted in any one building shall be of the same appearance and shall be cured by the same process. Units shall be free of any deleterious matter that will stain plaster or corrode metal. Concrete masonry units that have been subjected during manufacture to a saturated-steam pressure of not less than 120 pounds per square inch for 5 hours or more may be used as soon as cooled. Units that have not been subjected to such steam pressure shall be cured for 28 days or more before delivery. Units shall be delivered to the job site in an air-dry condition, and shall be classified into group 1 or group 2 depending upon the linear shrinkage potential as stipulated in table II.

*Table II. Linear Shrinkage Potential*

| Concrete density, lbs. per cu. ft. | Maximum linear shrinkage, percent, (ASTM Standard C 426) | |
|---|---|---|
| | Group 1 | Group 2 |
| 120 or more | 0.065 | 0.03 |
| Less than 120 | .065 | .04 |

(1) *Load-bearing concrete masonry units* conforming to ASTM Standard C 90, except as modified herein, shall be used in exterior walls and in interior load-bearing walls. Units may be either grade A or grade B, at the option of the contractor, except that only grade A units shall be used in uncoated and unpainted exterior walls. Units with face-shell thickness under 1¼ inches and over ¾ inches, with average minimum compressive strength of 1,000 pounds per square inch, and with no absorption requirement may be used in lieu of grade B units.

(2) *Non-load-bearing concrete masonry units* conforming to ASTM Standard C 129, except as modified herein, may be used in interior non-load-bearing walls and partitions, furring, and other non-load-bearing interior work.

(3) *Option:* The interior wythe of exterior non-load-bearing cavity walls may, at the option of the contractor, be constructed of either load-bearing or non-load-bearing units.

(4) *Solid load-bearing concrete masonry units* shall conform to ASTM Standard C 145, except as modified herein. Units may be either grade A or grade B, at the option of the contractor, except that only grade A units shall be used in uncoated and unpainted exterior walls.

(5) *Fire-resistant concrete masonry units* shall be of the type that will give fire rating required. Units shall be the rated product of a manufacturer listed in the current Fire

Protection Equipment List published by the Underwriters' Laboratories, Inc. In lieu of the above rating, fire-resistant units may be furnished on the basis of examination and report by a nationally recognized testing agency adequately equipped and competent to perform such services. The report shall state that the units furnished are equivalent in fire rating to those furnished by producers listed in the above Fire Protection Equipment List.

*g. Coping Tile.* Coping tile shall be salt-glazed fire-clay, terra cotta, or precast concrete units with socket joints. Units shall be sound, free from fractures, cracks, blisters, and warping, and shall be of standard size and proper width to overlap the wall masonry. Shapes required for external and internal angles shall be furnished.

*h. Firebrick.* Firebrick shall conform to Federal Specification HH–B–671, class 2, low-duty.

*i. Flashing Blocks.* Flashing blocks shall be hard-burned terra cotta with a diagonal groove not less than 1½ inches deep measured horizontally, to receive the flashing. The shapes shall provide a continuous groove around corners and for offsets. The blocks shall be of a size to replace and course with two courses of brick.

*j. Flue Lining and Thimbles.* Flue lining and thimbles shall be hard-burned fire clay or shale, shall be free from blisters and warping, and shall be of standard sizes and of sound manufacture. Flue linings of other materials and manufacturing processes having the approval of the Underwriters' Laboratories, Inc. or of any other recognized agency equally qualified to give such approval, may be used, subject to approval.

*k. Glazed Facing Tile Units.* Glazed structural facing-tile units for base shall be ceramic color-glazed or ceramic clear-glazed units as specified hereinafter, produced from a light-burning de-aired clay and conforming to ASTM Standard C 126, except as modified herein. Vertical external corners shall be bullnose. Vertical internal corners shall be square. Base shall be flush with the finished wall surface above and coved to meet finished floor surface only where ceramic mosaic tile flooring occurs. Ceramic color-glazed units shall have a satin finish with a (mottled) (smooth) texture and color range conforming to the approved samples. Ceramic clear-glazed units shall have a glossy finish with a translucent or tinted glaze in light cream or buff color and shall conform to the approved samples. On-the-job cutting of units shall be done with a motor-driven

saw designed for the purpose. Cut edges shall be clean, true, and sharp. Backs of units exposed in unfinished rooms and spaces shall be smooth and free from glaze. Backs of units upon or against which plaster is to be applied shall be scored, combed, or roughened, and shall be suitable for reception of plaster. Sides, ends, and backs of units upon or against which mortar is to be applied shall be reasonably free from glaze and suitable for the reception of mortar.

*l. Mortar.* Mortar shall comply with the property specification for type N mortar as set forth in ASTM Standard C 270 except that when tested for water retention the mortar shall have a flow, after suction of 75 percent or more when mixed to an initial flow of 125 to 140 percent. When tested for compressive strength, the water-retention requirements for mortar stipulated in ASTM Standard C 270 shall apply. When used in the work, mortar shall be mixed in the laboratory-established proportions with as much water as may be necessary to produce the workability desired regardless of initial flow. The contractor shall furnish a certified copy of laboratory-established proportions and tests as evidence that the mortar used in the work meets the requirements of the property specification as modified herein. No change in the laboratory-established proportions shall be made nor shall materials with different physical or chemical characteristics be utilized in mortar used in the work unless the contractor furnishes additional evidence that such mortar meets the requirements of the property specifications as specified herein.

*m. Precast Concrete Trim.* Precast concrete trim, unless otherwise shown, shall consist of class A concrete as specified in SECTION: CONCRETE, using ½-inch to No. 4 nominal-size coarse aggregate reinforced with not less than two No. 4 bars. Precast units shall have beds and joints at right angles to the face, with sharp true arrises, and shall have drip grooves on underside where units overhang the walls. Copings and sills shall be cast with washes, and where overhanging the walls shall have drips cut on the underside. Sills for windows having mullions shall be cast in sections with head joints at mullions and a ¼-inch allowance for mortar joints. The ends, except a ¾-inch-wide margin at exposed surfaces, shall be roughened for bond. Precast-trim items shall have an absorption of not more than 8 percent by weight after immersion in water for 48 hours. Unless precast-trim items have been subjected during manufacture to saturated-steam pressure of 120 pounds or more per square inch for

5 hours or more, the trim items shall, after casting, be damp-cured for 24 hours or more, or steam-treated and shall then be aged under cover for 28 days or longer. Prior to use, each item shall be wetted and inspected for crazing. Evidence of excessive crazing will be cause for rejection. Cast-concrete members weighing over 80 pounds shall have built-in loops of galvanized wire or other approved provisions for lifting and anchoring.

*n. Pre-Faced Concrete Units.* Pre-faced concrete masonry units may be used, at the option of the contractor, in lieu of glazed structural facing-tile units for base. Pre-faced concrete masonry units shall consist of concrete masonry units as specified herein, the finished in-place exposed-to-view surfaces of which are covered at the point of manufacture with a ceramic, a resin with sand filler, or a cement-bonded inseparable facing conforming to the requirements of ASTM Standard C 126 relative to imperviousness, opacity, and resistance to fading. Units tested for drying-shrinkage as specified shall be free from crazing. No face dimension of any facing shall vary from the corresponding face dimension of its concrete masonry unit by more than ¼ inch overall or by more than ⁵⁄₃₂ inch with respect to any one edge. Proposed units shall be tested at the expense of the contractor by an approved commercial testing laboratory for imperviousness, opacity, and resistance to fading. Certified reports of such tests shall be submitted to the Contracting Officer for approval prior to delivery of units to the project site. The facing shall be free from chips, cracks, crazes, blisters, crawling, holes, and other imperfections detracting from the appearance of the wall when viewed from a distance of 5 feet at right angles to the wall.

*o. Reinforcing Steel Bars and Rods.* Reinforcing steel bars and rods shall conform to Federal Specification QQ–S–632, type II, grade C, D, E or G, except that bars ¼ inch or less in diameter may be type II, grade A.

*p. Sand-Lime Brick.* Sand-lime brick conforming to ASTM Standard C 73 may be used, at the option of the contractor, in lieu of common-clay or shale brick as follows: grade SW or MW for back-up and interior work and grade SW for exterior facing and for work below grade. Sand-lime brick shall not be used for exterior facing unless such facing is noted as "sand-lime brick," on the drawings.

*q. Stone.* Stone for sills, lintels, and copings shall be cut to the design shown. Lintels, except when supported by a steel member, shall be 4 inches or more thick and of the depth required to support the masonry over the opening. All stone shall have

beds and joints at right angles to the face, with sharp, true arrises.

    (1) *Limestone* shall be standard buff limestone with a smooth machine finish. Exposed surfaces shall be free from tool marks.

    (2) *Sandstone* shall be standard grade, buff, gray, or buff-brown, with a smooth finish. Exposed surfaces shall be free from clay pits and tool marks.

    (3) *Granite* shall be a good commercial grade of building granite of medium or moderately coarse grain, and of a light or medium-gray or light-pink color, with smooth machine finish on wases, 4-cut finish on treads, and 6-cut or equivalent machine finish on other exposed surfaces.

    (4) *Terra cotta* manufactured in accordance with applicable specifications of the Architectural Terra Cotta Institute and of the color, texture, and finish selected by the Contracting Officer from samples submitted by the contractor may, at the option of the contractor, be used in lieu of stone for sills and copings.

*r. Wall Tile.* Structural clay wall tile of the sizes and shapes required may be used in lieu of concrete masonry units, at the option of the contractor. The exposed faces of the tile shall be smooth finished except where the tile serves as a base for stucco or plaster, or as setting beds for wall tile, in which locations the tile shall have a plaster-base finish. Smooth-faced structural clay wall tile shall be free from glaze, popouts, lime pits, and other disfiguring blemishes detracting from the appearance of the finished wall when viewed from a distance of 20 feet. Structural clay wall tile used in any one building shall be the same composition, size, and appearance. Structural clay wall tile shall be painted as specified in SECTION: PAINTING. In single-wythe two-faced walls or in partition walls, where such walls are in habitable rooms or spaces and are either to be painted or will be exposed to view, the tile shall not exceed 1½ percent difference in extreme dimensions. The tile shall include the closers, jamb, and other required shapes, and shall conform to the following:

    (1) *Load-bearing clay wall tile* shall be either vertical-cell or horizontal-cell type. Vertical-cell-type clay wall tile shall conform to ASTM Standard C 34. Horizontal-cell-type clay wall tile, shapes 9S, 16S, 17S, 19S, 20S, 21S, or 22S, as shown in Federal Specification SS–T–341, figure 1, with the

outer shell and adjacent vertical web not more than 1¼ inches apart, shall conform to ASTM Standard C 34. Units may be either grade LBX or LB, at the option of the contractor, except that only grade LBX units shall be used in uncoated and unpainted exterior walls.

(2) *Non-load-bearing clay wall tile* conforming to ASTM Standard C 56 shall be used in lieu of non-load-bearing concrete masonry units.

(3) *Fire-resistant clay wall tile* shall be of the type that will give the fire rating required, when subjected to the Standard Fire Tests of ASTM Standard E 119. A certified statement by a recognized testing laboratory indicating that the tile are capable of meeting the stipulated requirements shall be furnished the Contracting Officer with each shipment of tile.

(4) *Tile sills* for interior of windows shall be 6 by 6 by 12 inches and 6 by 8 by 12 inches, nominal dimensions, and shall conform to requirements for non-load-bearing structural clay wall tile.

## 4. Handling and Storage

Masonry materials shall be stored in an approved manner that will protect them from contact with soil and exposure to the elements.

## 5. Samples

The following samples of materials proposed for use shall be submitted to the Contracting Officer and his approval thereof received before materials represented by the samples are delivered to the project site. Representative samples shall be taken periodically from on-the-site stockpiles as required by the Contracting Officer for testing.

*a. Anchors and Ties.* Two of each type proposed for use.

*b. Clay or Shale Brick, Sand-Lime Brick, Concrete Masonry Units, Concrete Brick, Split Block, and Structural Clay Wall Tile.* All shapes, sizes, and kinds, in sufficient numbers to show full range of color and texture. After the samples are approved, sample panels shall be built on the project site where directed. Sample panels shall be 6 feet long, 4 feet high, and 4 inches thick, and shall show proposed color range, texture, bond, and mortar joint.

*c. Glazed Structural Facing-Tile Units for Base.* Shapes, sizes, and kinds in sufficient numbers to show full range of color and texture.

*d. Joint Reinforcement.* One piece of each type of reinforcement 18 inches long, showing at least two cross joints.

*e. Stone or Terra Cotta.* One piece approximately 6 by 8 inches in size, showing color, finish, and texture.

## 6. Certificates

The contractor shall furnish certificates executed in triplicate prior to delivery of the certified material to the project site. Each certificate shall be signed by an authorized officer of the manufacturing company and shall contain the name and address of the contractor, the project location, and the quantity and date or dates of shipment or delivery of the material to which the certificate applies. Concrete masonry units, concrete brick, split block, clay or shale brick, structural clay tile, precast concrete trim, and sandlime brick shall be certified for compliance with all specification requirements. Aggregate for concrete masonry units shall be certified for compliance with specification requirements for nonstaining and popout properties.

## 7. Tests for Drying-Shrinkage

Sampling and testing to determine the group classification of concrete masonry units, concrete brick, and split block shall be done by an approved commercial testing laboratory not more than 3 months nor less than 2 weeks before delivery of units to the project site, and at the expense of the contractor. Three copies of such tests shall be signed by the testing laboratory and countersigned by the contractor and shall be submitted to the Contracting Officer at least 10 days before delivery of units to the project site. No change in manufacturing processes and techniques or in drying and curing procedures shall be made nor shall materials with different physical or chemical characteristics be used in units delivered to the project site unless the contractor verifies the group classification by additional signed test reports.

*a. Samples for Testing.* A sample of 5 individual and whole units representative of the manufacturer's product whose units are proposed for use shall be selected after cooling and/or curing at the point of manufacture. Sample units shall prove under test to be free from cracks or other structural defects, and to have been manufactured with the same type and quality of aggregate, and cured and dried by the same procedures as those to be employed in producing; units for use in the work. Units previously subjected to tests involving temperatures exceeding 150° F. shall not be used in drying-shrinkage tests.

*b. Testing.* Testing shall be done in accordance with ASTM Standard C 426.

## 8. Test for Air-Dry Condition

Upon delivery of concrete masonry units, concrete brick, or split block to the project site, representative samples shall be selected from stockpiles and tested for air-dry condition. Sampling and testing will be by and at the expense of the Government and will be in accordance with ASTM Standard C 427. Air-dry condition is defined as the moisture condition of a concrete masonry unit, concrete brick, or split block in a state of equilibrium with a relative humidity of not greater than 15 percent higher than the average relative humidity at the project site, except that the relative humidity of the unit at equilibrium shall not exceed 85 percent and shall not be less than 50 percent in group I units nor less than 70 percent in group II units. The average relative humidity at the project site shall be as determined by the nearest US Weather Bureau station from the total of annual observations recorded for the month in which the unit is delivered.

## 9. Mortar Test

Test for mortar and to establish the proportions of the mortar to be used on the work shall be done by an approved commercial testing laboratory at the expense of the Government.

## 10. Erection

*a. General.* Masonry shall not be erected when the ambient temperature is below 35° F., except by written permission of the Contracting Officer. No frozen work shall be built upon. No brick or other unit having a film of water or frost on its surface shall be laid in the walls. Masonry shall be protected from freezing for 48 hours after being laid. Masonry shall be laid plumb, true to line, with level courses accurately spaced with a story pole, and, unless otherwise shown, with each course breaking joints with the course next below. Each unit shall be adjusted to its final position in the wall while mortar is still soft and plastic. Any unit that is disturbed after mortar has stiffened shall be removed and relaid with fresh mortar. Bond pattern shall be kept plumb throughout. Corners and reveals shall be plumb and true. Courses shall be so spaced that backing masonry will level off flush with the face work at all joints where metal ties are used. Free-standing load-bearing piers shall be bonded in each course. Walls or partitions abutting concrete columns or walls shall be anchored thereto with metal anchors or ties spaced not more than 16 inches on center vertically. Chases and raked-out joints shall be kept free from mortar or other debris. Spaces around metal door frames and other built-in items shall be solidly filled with mortar. Anchors, wall plugs, accessories, flashings, and other items required to be built in with masonry shall be built in as the masonry work progresses. Cutting and fitting of masonry required to accommodate the work of others shall be done by masonry mechanics with masonry saws. The sizes of any two adjacent units shall be within permitted tolerances so that the difference between the vertical faces of such units shall not exceed $\frac{1}{8}$ inch in exposed-to-view or painted walls and partitions in habitable rooms and spaces. Units in exposed-to-view or painted walls and partitions shall be free from chipped edges or other imperfections detracting from the appearance of the finished work.

*b. Joints.* Joints in exposed-to-view or painted walls and partitions, except joints to be calked, shall be tooled slightly concave with a device of as long length as practicable and so that the mortar will be thoroughly compacted and pressed against the edges of the units. Tooling shall not be done until after the mortar has taken its initial set. The following joints on the weather side of exterior masonry walls shall be raked out $\frac{3}{4}$ inch and left ready for calking:

    (1) Control joints.

    (2) Joints between metal frames and masonry.

    (3) Horizontal and vertical faces of joints at ends and in wash surfaces of slip sills and copings.

    (4) Other joints where so indicated.

*c. Unfinished Work.* Unfinished work shall be stepped back for joining with new work. Toothing may be resorted to only when specifically approved. All loose mortar shall be removed and the exposed joint shall be thoroughly cleaned before laying new work. Surfaces of masonry not being worked on shall be properly protected at all times during construction operations. When rain or snow is imminent and the work is discontinued, the tops of exposed masonry walls shall be covered with a strong waterproof membrane well secured in place. Adequate provisions shall be made during construction to prevent damage by wind.

*d. Mortar.* Mortar that has stiffened because of chemical reaction due to hydration shall not be used. Except as specified below, mortar shall be used and placed in final position within $2\frac{1}{2}$ hours after mixing where air temperature is 80° F. or higher, and within 3½ hours after mixing where air temperature is less than 80° F. Mortar not used within these time intervals shall be discarded. When cement or cements used in the mortar have been tested and the observed time of initial set as determined under ASTM Standard C 266 has been ascertained, the time interval during which the mortar must be placed in final position may be determined by an optional method as follows:

| Air temperature in degrees F. | Time interval after mixing |
|---|---|
| 80 or higher_____ | Time of initial set minus 1 hour |
| Less than 80_____ | Time of initial set minus ½ hour |

Mortars that have stiffened within the time interval as determined above, because of evaporation of moisture from the mortar, may be retempered by adding water as frequently as needed to restore workability.

*e. Brick.* Before being laid, clay or shale brick shall be wetted so as to have an initial rate of absorption of not more than 0.025 ounce per minute per square inch of bed surface, determined in accordance with ASTM Standard C 67. Concrete brick, split block, and sand-lime brick shall not be wetted before laying. Recessed brick shall be laid with the frog side down. Horizontal and vertical joints shall be completely filled with mortar when laid. Vertical joints shall be of the same width except for inconspicuous variations required to maintain the bond.

(1) *Brick facing.* Brick facing shall be laid up with the better face of the brick exposed. Unless otherwise indicated, brick facing shall be laid in running bond and shall be anchored or tied to the masonry backing with wire ties, wire-mesh ties, or corrugated or crimped-metal ties spaced not over 32 inches on centers horizontally, approximately 16 inches apart vertically, and extending not less than 3½ inches into the backing and to within ½ inch of the exposed surface of the facing joint. At concrete beams and columns, brick facings shall be anchored to the concrete by dovetail anchors set in grooves or inserts built in the face of concrete. The anchors shall be spaced not over 2 feet apart horizontally and 16 inches apart vertically in beams, and not over 16 inches apart vertically in center of columns. Shelf angles shall be adjusted as required to keep the masonry joints level and at the proper elevation.

(2) *Arches.* Arches shall be built over wood centering and supports provided as specified in SECTION: CARPENTRY. Supports and centering shall not be removed until so directed by the Contracting Officer.

(a) *Arches in face work,* except rowlock arches, shall have the brick ground or molded with the centerline of joints on radial lines. All radial joints shall be of equal width. Ground surfaces shall not be exposed. Arches in facing shall be bonded into or anchored to the backing. Flat arches shall be constructed with a camber of 1/16 inch per foot of span.

(b) *Rowlock arches* shall be turned over all openings in interior bearing walls and fire walls that are so indicated.

(3) *Special shapes.* Special molded or ground-brick shapes shall be used for ornamental work where required.

(4) *Joints.* Exposed mortar head and bed joints shall have a thickness equal to the difference between the actual and nominal dimensions of the brick in either height or length, but in no case shall the average width of any three adjacent joints be less than ¼ or more than ½ inch.

(5) *Back parging.* In brick-faced masonry-backup walls, the faces of masonry units back of brick facing or the backs of the facing bricks shall be parged with mortar as the wall is built. Parging shall be not less than ½ inch thick troweled to a smooth dense surface. Parging of cavity walls will not be required.

(6) *Cavity-wall construction.* In walls of cavity construction, unless otherwise indicated, the facing and backing masonry shall be completely separated by a continuous air space not less than 2 nor more than 3 inches wide, except for returns at jambs of openings. The two wythes shall be securely tied together by cavity-wall ties staggered and spaced not to exceed 36 inches apart horizontally and 16 inches apart vertically. Additional cavity-wall ties shall be placed within 8 inches of the jambs of all openings except where the wythes are bonded together with masonry returns at jambs. The air space between

the facing and backing wythes shall be kept clear and clean of mortar droppings by temporary wood strips laid on the wall ties and carefully lifted out before the next row of ties or anchors is placed.

(a) *Facing over columns and spandrel beams* shall be tied thereto with anchors having one end formed to fit into dovetail channels set into the concrete. One line of ties shall be installed just above the roof slab where parapet walls occur, and in the mortar joint under spandrel beams supporting the roof slab.

(b) *Weep holes* shall be provided 24 inches apart on centers in mortar joints of exterior wythe along the bottom of the cavity over foundations, bond beams, and other water stops in wall, by placing short lengths of well-greased No. 10, $\frac{5}{16}$-inch nominal diameter, braided cotton sash cord in the mortar and withdrawing these pieces of cord after the wall has been completed. Other methods of providing such weep holes may be used subject to approval.

(7) *Chimneys.* Chimneys, unless otherwise indicated, shall be built of brick, lined with fire clay or other approved flue lining of the size indicated. Flue lining shall extend from 1 foot below the smoke inlet to the full height of the chimney and 2 inches or more above the chimney cap. Thimbles shall be placed as indicated or as directed. The space between the lining and the enclosing masonry shall be filled solid with mortar. Linings shall be built in as the work progresses. The vertical joints of top course of brickwork of chimneys shall be raked out about $\frac{3}{4}$ inch deep to provide a key for the cement-mortar setting bed for the chimney cap or cement-mortar wash on top of brickwork. Where not otherwise indicated, chimney tops shall be provided with a wash-type cap, 1 inch or more thick at its outer edge, composed of 1 part portland cement and 2 parts well-graded coarse sand and reinforced with two rings of No. 3 gage or heavier galvanized steel wire having ends lapped 6 inches.

(8) *Fireplaces.* Fireplaces, unless otherwise indicated, shall be faced with the same type of brick that is used for the facing of exposed-exterior masonry walls. Backing construction of fireplaces shall have built-in metal ties 1 foot on centers in every other course, to tie in the fireplace facing and lining. Fireplaces shall be lined on the back and sides of the fire chamber with firebrick selected for uniformity of shape and color. Back hearth in fire chamber shall also be of firebrick when so indicated. Firebrick shall be laid with mortar joints not more than $\frac{1}{4}$ inch wide. Hearth brick, when indicated, shall be of a reasonably smooth surface of approximately the same color as the fireplace facing, and shall be laid on a full bed of mortar, with flush joints in the pattern indicated. Metal damper, angle lintel, and ash dump shall be built into the adjoining brickwork with all spaces between the metal items and brickwork filled solid with mortar.

f. *Concrete Masonry Units and Structural Clay Tile.* Structural clay tile, before being laid, shall be wetted so as to have an initial rate of absorption of not more than 12 percent when determined in accordance with ASTM Standard C 112. Concrete masonry units shall not be wetted before laying. Cutting of units shall be accomplished by masonry mechanics using masonry saws. Concrete units shall be dry cut. Units shall be set with vertical joints breaking not less than 4 inches over units in the course next below. Mortar joints shall be approximately $\frac{3}{8}$ inch wide. Mortar joints in piers, columns, and pilasters, and starting courses on footings, on solid foundation walls, and on beams shall be full bedded under both face shells and webs. Other joints shall have full mortar coverage on horizontal- and vertical-face shells, but mortar shall not extend through the unit on the web edges. Each course shall be bonded at corners. Jamb units shall be of the shapes and sizes required to bond with wall units. No cells shall be left open in face surfaces. Sections of brickwork shall be incorporated in the masonry work where necessary to fill out at corners, gable slopes, and elsewhere as required. Masonry-unit walls or partitions supporting plumbing, heating, or other fixtures, and voids at door and window jambs, and other spaces requiring grout fill shall be full bedded in mortar to prevent leakage and filled solid with mortar mixed to pouring consistency.

(1) *Parging.* The face or faces of exterior concrete-masonry-unit walls below grade against which backfill is to be placed shall be parged with mortar. Parging shall be not less than $\frac{1}{2}$ inch thick, trowled to a

smooth dense surface so as to provide a continuous unbroken shield from top of footings to a line 6 inches above adjacent finish grade, unless otherwise indicated. Parging shall be damp cured for 48 hours or more before fill or backfill is placed thereagainst.

(2) *Partitions.* Partitions along corridors, at stairways, at boiler and heater rooms, and at pipe spaces and elevator shafts shall be continuous from floor to underside of floor or roof construction above. Where suspended ceilings on both sides of partitions are indicated, the partitions in spaces other than those mentioned above may be stopped approximately 4 inches above the ceiling level, unless otherwise indicated.

(3) *Lintels.* Lintels in masonry-unit partitions and furring, unless otherwise indicated, shall be constructed of specially formed lintel- or U-shaped units filled with class-B concrete as specified under SECTION: CONCRETE, using coarse aggregate of ½-inch to No. 4 nominal size, and shall be reinforced as indicated. However, not less than two No. 4 bars the full length of the lintel shall be provided. Lintels shall extend at least 8 inches beyond each side of the opening. The bed joints of lintels at control joints shall be underlaid with a sheet of 16-ounce smooth copper with edges cut back ½ inch from face of wall below.

(4) *Furring.* Furring, when free-standing, shall be built of hollow partition units and, unless otherwise indicated, shall be 3 inches thick, laid with cells vertical except at lintels and window sills. Unless otherwise indicated, the furring units shall be returned on the jambs and other reveals, and around projections that are indicated on the drawings as furred with masonry. Furring shall be suitably anchored to masonry walls or concrete with built-in metal anchors spaced not over 2 feet on centers horizontally and vertically and placed so as to coincide with the joints in masonry furring. Split-unit furring 2 inches thick may be used only when the furring is built in direct contact with the surfaces of wall masonry or concrete. Masonry furring less than 3 inches thick shall be anchored with metal ties spaced as specified above for free-standing furring. Untied furring

shall be spot bedded in mortar against the backing masonry. Furring at columns and for pipe-space enclosures shall not be bonded to other masonry but shall be anchored thereto by metal ties.

(5) *Fireproofing.* Fireproofing around steel columns shall be 4 inches or more thick, shall be set at least ½ inch away from the steel, shall not be cut or reduced in thickness at splice plates, and shall not be omitted or reduced in thickness at abutting pipe spaces.

(6) *Cavity-wall construction.* Unless otherwise indicated, the inner and outer wythes of cavity walls shall be completely separated by a continuous air space not less than 2 inches nor more than 3 inches wide, except for masonry returns indicated at jambs of openings. The two wythes shall be securely tied together by cavity-wall ties staggered and spaced not to exceed 36 inches apart horizontally and 16 inches apart vertically. Additional cavity-wall ties shall be positioned within 8 inches of the jambs of openings and not more than 2 feet apart vertically, except where the wythes are bonded together with masonry returns at jambs. The inner and outer wythes of cavity walls shall be provided with control joints and joint reinforcement as hereinafter specified for single-thickness exterior walls. At control joints, the alinement of both wythes shall be maintained by ties located 16 inches apart on centers vertically along each side of the control joint. The air space between the facing and backing wythes shall be kept clear and clean of mortar droppings by temporary wood strips laid on the wall ties and carefully lifted out before the next row of ties or anchors is placed.

(a) *Facing over columns and spandrel beams* shall be tied thereto with anchors having one end formed to fit into dovetail channels set into the concrete.

(b) *Weep holes* shall be provided 32 inches on centers in mortar joints of the exterior wythe along the bottom of the cavity over foundations, bond beams, and other water stops in the wall. The holes may be formed by placing short lengths of well-greased No. 10, 5⁄16-inch nominal diameter, braided cotton sash cord in the mortar and withdrawing these pieces of cord after the wall has been completed. Other methods of providing weep holes may be used subject to approval.

(7) *Vertical cells.* Vertical cells to be filled shall have vertical alinement sufficient to maintain a clear, unobstructed, continuous vertical cell measuring not less than 2 by 3 inches. Vertical reinforcement shall be continuous and rigidly secured at top and bottom and at intervals necessary to hold the reinforcing in proper position.

(8) *Reinforcement units.* Units containing reinforcement shall be solid filled with class-B concrete, as specified under SECTION: CONCRETE, using ½-inch to No. 4 nominal-size coarse aggregate.

*g. Glazed Facing Tile Units.* Glazed structural facing-tile units for bases shall be set level and true to line in a full bed or mortar with vertical joints filled. Faces of units shall be kept free of mortar. No piece shorter than 4 inches shall be used at any vertical corner or jamb.

*h. Sills, Lintels, and Copings.* Sills, lintels, and copings shall be set with faces plumb and true, in a full bed of mortar, except that sills with lugs shall have mortar beds under the ends of the sill only. Sills shall be leveled and tapped into place on these beds. Upon completion of the walls the remainder of the bed joint shall be filled solidly with mortar from front and back, and the exterior face of the mortar tooled smooth.

## 11. Shrinkage-Cracking Control

Shrinkage cracking in concrete-masonry-unit and concrete-brick and split-block construction shall be controlled by combinations of control joints, joint reinforcing, and bond beams in strict accordance with the details indicated and as specified hereinafter.

*a. Bond beams* shall consist of units filled with concrete and reinforced as indicated. Concrete filling shall be class B as specified under SECTION: CONCRETE. Bond beams shall be broken at expansion joints and, where indicated, at control joints. Dummy control joints shall be formed in the bond beam where bond beam is not broken at control joint.

*b. Joint reinforcement* shall be installed as indicated. Joint reinforcement at openings shall extend not less than 24 inches beyond the end of sills and lintels or to the end of the panel if the distance to the end of the panel is less than 24 inches. Reinforce-

ment shall be lapped 6 inches or more, and the lap shall contain at least one cross wire of each piece of reinforcement. Joint reinforcement shall be accurately formed around corners at wall intersections.

*c. Control joints* shall be provided at the locations indicated and shall be constructed by using either special control-joint units or open-end stretcher units, at the option of the contractor. On the exterior face of the wall, the control joint shall be raked to a depth of ¾ inch and left ready for calking. Control joints on exposed-to-view or painted interior walls shall be raked to a depth of ¼ inch and shall not be calked. Calking of control joints is covered under SECTION: CALKING, and is specified therein.

## 12. Expansion Joints

*a. Location.* Expansion joints shall be located where indicated.

*b. Size.* Joints shall be not less than 1½ inches nor more than 2 inches wide.

*c. Installation.* Method of installation shall be as follows, unless otherwise indicated. Units on each side of joint shall be steel-sash jamb units having a ¾- by ¾-inch groove near the center at end of each unit. A weather seal consisting of flexible joint-filler strip ½ inch thick and of sufficient length to reach continuously the entire story height shall be set into the grooves of units with plastic calking compound. The weather side of the joint shall be covered with a sheet-metal bellows strip suitably secured to the wall with corrosion-resistant fasteners. The interior side of the joint shall be covered with a sliding plate of not lighter than 14-gage steel, prime-coated for painting, secured with suitable fasteners to one side only of the wall. Sheet-metal bellows strip is covered under SECTION: SHEET METAL, and is specified therein.

## 13. Pointing and Cleaning

Before completion of the work, all defects in joints of exposed exterior masonry surfaces shall be raked out as necessary, filled with mortar, and retooled. After pointing mortar has set and hardened, all exposed clay-brick and clay-tile-masonry surfaces shall be wetted and then cleaned with a solution of 10 percent by volume of commercial muriatic acid applied with stiff-fiber brushes, and immediately after cleaning, the surfaces shall be thoroughly rinsed down with clear water. All masonry surfaces shall be left clean, free of mortar daubs, and with tight mortar joints throughout.

## 14. Payment

No separate payment will be made for the work covered under this section, and all costs in connection therewith shall be included in the lump-sum contract price for the structure to which the work pertains.

## NOTES TO CONTRACTING OFFICER

1. This specification is to be used in the preparation of contract specifications. It will not form a part of a contract merely by reference; pertinent portions will be copied verbatim into the contract documents.

2. The section number will be inserted in the specification heading and prefixed to each paragraph and page number.

3. Paragraph 2: The listed designations for publications are those that were in effect when the specification was being prepared. These designations will be changed, as required, to those in effect on the date of invitation for bids; and the nomenclature, types, grades, classes, etc., referenced in the specifications will be checked for conformance to the latest revision or amendment. To minimize the possibility of error, the letter suffixes, amendments, and dates indicating specific issues will be omitted elsewhere in the specification. It is essential, therefore, that the list of applicable publications be retained in the contract specifications.

4. Paragraph 3a: Where sufficient evidence based on previous construction experience indicates concrete masonry units manufactured from aggregate from a specific source may be subject to excessive popouts and/or staining, contract specifications may be written to exclude such aggregate.

5. Paragraphs 3f(5) and 3r(3): The fire resistance classifications of walls constructed of concrete masonry units and structural clay tile are given in tables 23 and 20, respectively, of NBS Report BMS 92, Fire-Resistance Classifications of Building Construction.

6. Paragraphs 3k and 3n: Where projects require glazed structural facing-tile units at locations other than for bases, these paragraphs will be deleted from the MASONRY section of contract specifications and included in a STRUCTURAL FACING UNITS section based on Applicable Specification.

7. Paragraphs 3l and 10d: This specification contains the mortar requirements for masonry work specified herein; therefore other Specification for MORTARS, MASONRY will not be used. Requirements for structures in seismic areas or for wind loads in excess of 20 pounds per square foot of wall surface are not within the scope of this specification, and

mortars with higher than 750 psi compressive strength are not included for this reason.

8. Paragraph 3m: Cast-in-place trim will not be used. Sills will be precast concrete. Lintels may be either precast concrete or may be constructed of specially formed lintel blocks. Bond beams will be formed from special units.

9. Paragraph 3q: Stone will be included only in cases where the structure requires a limited quantity of cut stone steps, sills, lintels, copings, or other minor trim items that can be set by the brick mason. Where copings are included, this paragraph will be modified to include the necessary jointing, calking, and anchorage requirements from Specification on STONEWORK. In all other cases use a separate SECTION for STONEWORK. Similarly, where rubble-stone foundation walls, facing for retaining walls, steps, and other site-improvement work are required, a separate SECTION: STONE MASONRY, will be used to cover that work. Stone sills, lintels, and copings, if required, will be indicated, and paragraphs 3q(1), (2), and/or (3) included in the contract specifications as desired, and the option for terra cotta may be included at the discretion of the Contracting Officer.

10. Paragraph 3r: It is mandatory that the option for structural clay wall tile be incorporated into the contract specifications. Unglazed, unpainted structural clay facing tile will also be incorporated into the contract specifications as a contractor's option for painted concrete masonry units if the color of the units does not create an esthetic problem. Where so incorporated, ASTM Standard C 212 will be inserted in the APPLICABLE PUBLICATIONS paragraph, and the following will be added to paragraph 3:

> s. *Structural clay facing tile* conforming to ASTM Standard C 212, type FTS or FTX, may be provided at the option of the contractor in lieu of painted concrete masonry units. The range of color shall be as selected by the Contracting Officer from samples submitted for approval. Solid-shell and cored-shell units of side construction shall not be used.

11. Paragraph 5: For small buildings and where appearance is not of major importance, the requirement for sample panels showing color, bond, mortar joints, and texture may be omitted. The other requirements for samples, except concrete masonry units, concrete brick, and sand-lime brick, may be modified at the discretion of the Contracting Officer.

12. Paragraph 8: Copies of referenced bulletin may be obtained upon request from the Portland Cement Association, 33 West Grand Avenue, Chicago 10, Ill.

13. Paragraph 10: In the event the duration of the construction contract will require cold-weather placement of masonry, the following subparagraphs will be inserted after subparagraph *a* and the subsequent subparagraphs of paragraph 10 is redesignated accordingly:

*b. Cold-weather installation:* When masonry work is authorized during temperatures below 35° F., special protective provisions as follows shall be provided:

   (1) When the outside air temperature is between 35° F. and 32° F., all masonry units shall be kept completely covered and free from ice and snow at all times. Either the mixing water or sand for mortar shall be heated to a temperature between 70° F. and 160° F. The air temperature on both sides of the masonry shall be maintained above 40° F. for a period of at least 72 hours. The contractor shall submit for approval a written statement of the methods he proposes to use for protecting the masonry against low temperatures. Building upon frozen work is prohibited.

   (2) When the outside air temperature is between 32° F. and 25° F., in addition to the above provisions, both the mixing water and sand shall be heated to a temperature between 70° F. and 160° F.

   (3) When the outside air temperature is between 25° F. and 18° F., in addition to the above provisions, calcium chloride shall be added to the mixing water at a rate not to exceed 1½ to 2 pounds per sack of portland cement. Calcium chloride shall not exceed 1 percent for masonry cement

mortars unless specifically recommended by the manufacturer of the masonry cement.

   (4) When the outside air temperature is between 18° F. and 0° F., in addition to the above provisions, all masonry units when laid shall be heated to at least 40° F.

   (5) When the outside air temperature is 0° F. and below, masonry work shall be performed only in emergency construction. In addition to the above provisions, all masonry units shall be heated to a temperature of at least 40° F. Complete temporary inclosures shall be provided for masonry construction during sustained subzero weather.

14. Paragraph 10*e*(7): An option for radial-block chimneys, will be included in contract specifications where brick chimneys 60 feet or more in height are required. Stub chimneys with induced-draft fans will be designed where required by the using service. Chimneys having flue openings of 576 square inches or less and not over 40 feet high may, at the discretion of the Contracting Officer, be constructed of prefabricated units that comply with National Board of Fire Underwriters' requirements.

15. Paragraph 10*f*(2): The minimum thickness of partitions enclosing pipe spaces, elevator, and stair shafts will be determined in accordance with the latest edition of the recommended building code of the National Board of Fire Underwriters, unless the applicable State or local building code has more rigid requirements, in which case the more restrictive provisions will govern.

16. Paragraph 10*f*(5): For fireproofing around structural steel columns, the fire-resistance ratings required for both exterior wall columns and interior construction will be determined from design criteria as established by the latest edition of the Recommended Building Code of the National Board of Fire Underwriters unless the applicable State or local building code has more rigid requirements, in which case the more restrictive provisions will govern. Where masonry is to be used for fireproofing, the thickness of the tile or concrete masonry units to be used will be checked against the fire-resistance rating

required, and the minimum thickness of covering specified will be revised, when necessary, to conform to the minimum thickness of fireproof covering for the fire resistance called for. Other kinds of masonry or combination masonry and other materials may be substituted for structural clay tile as follows: the total thickness shall include the combined thickness of the individual materials based upon the protective fire-resistance ratings of the different materials to be used,

as established in the Code of National Board of Fire Underwriters referenced above.

17. Paragraph 11: For concrete masonry the locations and details of bond beams, control joints, and joint reinforcement for shrinkage-cracking control will be shown on contract drawings, as outlined in current design directives.

18. Paragraph 14: The PAYMENT paragraph will be deleted from any specification contemplating one lump-sum price for the entire work covered by the invitation for bids.

# APPENDIX IV

## SPECIFICATION FOR WATERPROOFING, BITUMINOUS-MEMBRANE

### 1. Scope

The work covered by this section of the specifications consists in furnishing all plant, labor equipment, appliances, and materials, and in performing all operations in connection with the installation of bituminous-membrane waterproofing, complete, in strict accordance with this section of the specifications and the applicable drawings, and subject to terms and conditions of the contract.

### 2. Applicable Specifications

The following publications of the issues listed below, but referred to thereafter by basic designation, form a part of this specification to the extent indicated by references thereto:

*a. Federal Specifications:*

| | |
|---|---|
| R–P–381 & Am–1 | Pitch; Coal-Tar (for) Mineral-Surfaced Built-Up Roofing, Waterproofing, and Damp-Proofing. |
| HH–C–581a | Cotton Fabric; Woven, Asphalt-Saturated. |
| HH–C–591 | Cotton Fabric; Woven, Coal-Tar-Saturated. |
| HH–F–191a & Am–2 | Felt; Asphalt-Saturated (for) Flashings, Roofing, and Waterproofing. |
| HH–F–201 & Am–1 | Felt; Coal-Tar-Saturated (for) Roofing and Waterproofing. |
| SS–A–666 & Am–1 | Asphalt; (for) Built-Up Roofing, Waterproofing, and Damp-Proofing. |
| SS–A–701 | Asphalt-Primer; (for) Roofing and Waterproofing. |
| LLL–F–321b & Am–1 | Fiberboard; Insulating. |

*b. American Society for Testing and Materials Standard:*

| | |
|---|---|
| D 1327–5T | Woven Burlap Fabrics Saturated with Bituminous Substances for Use in Waterproofing. (Tentative) |

### 3. General

Waterproofing shall be either asphalt or tar at the option of the contractor and shall be installed at the locations shown on the drawings. For walls of subgrade spaces and for floors on grade, waterproofing shall be 5-ply. For floors other than those on grade, waterproofing shall be 3-ply. Waterproofing shall not be applied when the ambient temperature is 40° F. or below. Waterproofing on floor slabs above grade shall be turned up around the walls or partitions inclosing such areas for a height of at least 4 inches, unless other height is indicated on the drawings. Where pipes pass through floor areas to be waterproofed, or where floor drains occur in such areas, the waterproofing shall not be installed until after the flashing around the pipes and drains has been installed. Flashing thus installed shall be lapped into the plies of the waterproofing and mopped in thereto in a manner that will assure a waterproof joint.

### 4. Materials

All materials shall be delivered to the site in sealed containers bearing the manufacturers' original labels.

*a. Asphalt* shall conform to Federal Specifications SS–A–666, type III.

*b. Asphalt primer* shall conform to Federal Specifications SS–A–701.

*c. Asphalt-saturated fabric* shall conform to Federal Specification HH–C–581. At the option of the contractor, either thermoplastic-bitumen-treated fibrous-glass membrane conforming to the appearance, average strength, and pliability requirements of Federal Specification HH–C–581 or asphalt-saturated woven burlap fabric conforming to ASTM Specification D 1327 may be used in lieu of asphalt-saturated fabric.

*d. Asphalt-saturated felt* shall conform to Federal Specification HH–F–191, type I.

*e. Coal-tar pitch* shall conform to Federal Specification R–P–381, type II.

*f. Coal-tar-saturated fabric* shall conform to Federal Specification HH–C–591. At the option of the contractor, either thermoplastic-bitumen-treated fibrous-glass membrane conforming to the appearance, average strength, and pliability requirements of Federal Specification HH–C–591 or coal-tar-saturated woven burlap fabric conforming to ASTM Specification D 1327 may be used in lieu of coal-tar-saturated fabric.

*g. Coal-tar saturated felt* shall conform to Federal Specification HH–F–201, type I.

*h. Insulating fiberboard* shall conform to Federal

Specification LLL-F–321, class A, ½ inch thick, and shall be treated to prevent destruction by termites.

## 5. Samples and Descriptive Data

Before delivery of any material to the site, the following samples or descriptive data, as indicated, shall be submitted for approval:

Asphalt-saturated or coal-tar-saturated fabric, 3 pieces, size 12 inches x width of roll.

Asphalt-saturated or coal-tar-saturated felt, 3 pieces, size 12 inches x width of roll.

Insulating fiberboard, descriptive data.

## 6. Preparation of Surfaces

Surfaces to receive waterproofing shall be clean and dry. All holes, joints, and cracks shall be pointed flush with mortar, and high spots shall be cut off or ground smooth. Before waterproofing is applied, the surfaces to be covered shall be carefully swept to remove all dust and foreign matter, and shall have been inspected and approved by the Contracting Officer.

## 7. Application of Waterproofing

For tar-waterproofing, surfaces shall be coated with a hot mopping of coal-tar pitch into which the required number of plies of coal-tar-saturated felt or fabric shall be embedded. For asphalt-type waterproofing, surfaces shall be given a coat of asphalt primer which shall be allowed to dry and then shall be coated with a hot mopping of asphalt into which the required number of plies of asphalt-saturated felt or fabric shall be embedded. The pitch or asphalt shall be heated to flow freely, but the pitch shall not be heated above 400° F., nor the asphalt above 450° F. The pitch or asphalt coatings shall be hot when the felt or fabric is embedded therein. The felt or fabric shall be applied so as to be free from wrinkles or buckles and shall be completely covered by a coating of pitch or asphalt in such manner as to separate completely each ply from the underlying ply over the entire area. The entire top surface shall then be given a final mopping, using not less than 70 pounds of pitch, or 60 pounds of asphalt, per 100 square feet.

a. *Type of Membrane.* Either felt or fabric may be used at the option of the contractor except where fabric-membrane waterproofing is indicated on the drawings or required. Where waterproofing is shown as carried through foundation walls that are keyed to the footing, the portion of the membrane extending through the foundation walls, and an adequate lap on both sides for bonding into the adjacent membrane on the exterior of the wall and in the floor, shall be of the saturated fabric. Where waterproof membrane on the floor is turned up at walls, at vertical angles in walls, and at any other places where the waterproof membrane may be subjected to unusual strain, strips consisting of two additional plies of saturated fabric and alternate moppings of asphalt or pitch shall be applied. Such strips at floors and wall angles shall be of sufficient width to extend at least 6 inches on the floor and 4 inches up the wall. Strips at vertical corners shall extend at least 5 inches on each side of the corner.

b. *Three-Ply Membrane.* Three-ply membrane shall be applied by the shingle method with each strip lapped over the preceding strip 22 inches if 32-inch-wide material is used, or 27½ inches if 36-inch-wide material is used.

c. *Five-Ply Membrane.* Five-ply membrane shall be applied by the shingle method with each strip lapped over the preceding strip 26 inches if 32-inch-wide material is used or lapped 29 inches if 36-inch-wide material is used; or by the 2- and 3-ply method, having two plies with each strip lapped over the preceding strip 17 inches if 32-inch-wide material is used or lapped 19 inches if 36-inch-wide material is used, followed by three plies lapped as specified above for 3-ply membrane work.

## 8. Protection

Waterproofing applied to walls against which backfill is to be placed shall be protected by a single thickness of insulating fiberboard. The insulating fiberboard shall be pressed into the final mopping while the mopping is still hot, with edges of boards brought into moderate contact and joints staggered. Boards shall be carefully and neatly fitted around pipes and projections and shall cover the entire surface of the waterproofing. Membranes that are not covered with insulating fiberboards shall be given temporary protection to prevent injury to the membrane by subsequent building operations.

## 9. Payment

No separate payment will be made for the work covered under this section, and all costs in connection therewith shall be included in the lump-sum contract price for the entire work to be performed under this contract.

# NOTES TO CONTRACTING OFFICER

1. This specification is to be used in the preparation of contract specifications in accordance with current instructions.

2. The section number should be inserted in the specification heading and prefixed to the paragraph and page numbers, and should be inserted where any of the paragraphs are referenced.

3. Paragraph 2: The listed designations for specifications and standards are those that were in effect when the guide specification was being prepared. These designations should be changed, as required, to those in effect on the date of invitation for bids, and the nomenclature, types, grades, and classes, referenced in the guide should be checked for conformance to the latest revision or amendment. To minimize the possibility of error, the letter suffixes or dates indicating specific issues and amendments should be omitted elsewhere in the specification. It is essential, therefore, that the list of applicable publications be retained in the contract specifications.

4. The location and extent of waterproofing should be indicated clearly on the drawings, or paragraph 3 should be revised to describe areas to be waterproofed.

5. In well-drained soils where no head of water is present, the use of drain tile and dampproofing will usually be sufficient, and waterproofing of such subgrade spaces for protection against rain and outside moisture only should not be required.

6. Floors in shower rooms, lavatories, hydrotherapy-treatment rooms, dishwashing rooms, laundries, and other spaces subject to continuous splashing of water, if over finished spaces, should be waterproofed.

7. Paragraph 7a. In cases where waterproofing must be applied to concrete or masonry walls in waterlogged soils or where some settlement is likely to occur; it may be advisable to use the fabric type instead of the felt type. In cases where rough masonry walls must be waterproofed, unless such walls can be made reasonably smooth with a parging of cement mortar, the fabric type of membrane waterproofing only should be specified.

# APPENDIX V

## SPECIFICATION FOR DAMPPROOFING

### 1. Scope

This section covers dampproofing, complete.

### 2. Applicable Publications

The following publications of the issues listed below, but referred to thereafter by basic designation, form a part of this specification to the extent indicated by references thereto:

*a. Federal Specifications:*

R-P-381 & Am-1   Pitch; Coal-Tar (for) Mineral-Surfaced Built-Up Roofing, Waterproofing, and Damp-proofing.

SS-A-666 & Am-1   Asphalt; (for) Built-Up Roofing, Waterproofing, and Damp-proofing.

SS-A-701   Asphalt-Primer; (for) Roofing and Waterproofing.

SS-R-451   Roof-Coating; Asphalt, Brushing-Consistency.

TT-C-655   Creosote, Technical, Wood Preservative, (for) Brush, Spray, or Open-Tank Treatment.

*b. American Society for Testing and Materials Standard:*

D 1187-56   Asphalt-Base Emulsions for Use as Protective Coatings for Metal.

### 3. General

Dampproofing shall not be applied when the temperature is below 40° F. and falling. The work shall be done by workmen experienced in the application of dampproofing, and the contractor shall coordinate dampproofing operations with other phases of the work to prevent staining or damaging finished work. The contractor shall repair or replace damaged finished work to the satisfaction of the Contracting Officer and without additional cost to the Government.

*a. Exterior Surfaces.* Dampproofing shall be applied to exterior surfaces of walls enclosing excavated basement and other subsurface spaces indicated, and shall extend from the top of the footings to within 4 inches of the finished grade. Unless otherwise indicated, walls having brick or stone facing backed by concrete and extending below grade shall have dampproofing applied to the con-crete backing, up to a line one foot above the finished grade or to the top of the concrete where the concrete does not extend one foot above the grade.

*b. Interior Surfaces.* Dampproofing shall be applied to the interior surfaces of all exterior masonry walls other than cavity-type walls except where wall surfaces are to be the finished walls of rooms. The interior surfaces of exterior non-cavity-type masonry walls in attic spaces and other unfinished spaces shall be dampproofed. Dampproofing shall be applied to the jambs and heads of openings and to all recessed spaces and shall extend not less than 1 foot in from the exterior walls on all interior masonry or concrete walls or partitions except where the surfaces of such walls or partitions are to be the finished walls of rooms. Wall surfaces back of shower linings and other surfaces so indicated shall be coated with interior dampproofing.

*c. Concrete Surfaces Back of Stone Facing.* Dampproofing shall be applied to all concrete surfaces against which stone is to be installed without intervening material other than mortar. Where such concrete is a pier, column, or beam, the damp-proofing shall return at least 6 inches on the sides of the concrete.

### 4. Materials

Materials shall be delivered in sealed containers bearing the manufacturer's original labels. The following materials shall conform to the respective specifications designated below:

*a. Asphalt.* Federal Specification SS-A-666, type III.

*b. Asphalt-Base Emulsions.* ASTM Standard D 1187.

*c. Asphalt Primer.* Federal Specification SS-A-701.

*d. Coal-Tar Pitch.* Federal Specification R-P-381, type II.

*e. Creosote Oil.* Federal Specification TT-C-655.

*f. Fibrous Asphalt.* Shall conform to the requirements of Federal Specifications SS-R-451, except that it shall contain not less than 12 percent of asbestos fiber.

### 5. Samples

Before delivery of any material to the site, the

following samples shall be submitted to the Contracting Officer for testing and approval:

| | |
|---|---|
| Asphalt | 5 pounds |
| Asphalt-base emulsion | 1 quart |
| Asphalt primer | 1 quart |
| Coal-tar pitch | 5 pounds |
| Creosote oil | 1 quart |
| Fibrous asphalt | 1 quart |

## 6. Preparation of Surfaces

Surfaces to receive dampproofing shall be smooth, clean, and dry. Holes, joints, and cracks shall be pointed flush with mortar and high spots ground level with the surrounding surface. Before dampproofing, surfaces shall be swept clean of all foreign matter and shall be inspected and approved.

## 7. Exterior Dampproofing

*a. Hot-Application Type.* Exterior surfaces to be dampproofed shall be given either a priming coat of creosote oil and two mop coats of hot coal-tar pitch or a priming coat of asphalt primer and two mop coats of hot asphalt. The mop coats shall be applied uniformly, using not less than 25 pounds per 100 square feet per coat. Pitch or asphalt shall be heated to flow freely but not to more than 375° F. for coal-tar pitch and 400° F. for asphalt. The finished surface shall be smooth, lustrous, and impervious to moisture. Dull or porous spots shall be recoated.

*b. Cold-Application Type.* Exterior surfaces to be dampproofed shall be given a priming coat of asphalt primer and a coat of fibrous asphalt applied uniformly, using not less than 30 pounds per 100 square feet, in such manner as to cover all pores completely and to thoroughly bond with the wall surface.

## 8. Interior Dampproofing

Interior surfaces indicated to be dampproofed and surfaces back of stone facing shall be given a coat of asphalt primer and a heavy coat of fibrous asphalt applied uniformly, using not less than 30 pounds per 100 square feet, in such manner as to cover all pores completely and to thoroughly bond with the wall surface, or at the contractor's option, may be given a priming coat and second coat of asphalt-base emulsion. The priming coat shall be applied in accordance with the manufacturer's printed directions at the rate of 1 gallon per 100 square feet and the second coat at the rate of 3 gallons per 100 square feet of surface. Interior dampproofing shall be applied before wall furring or furring strips are installed. The coating shall be free from runs and sags at normal temperatures.

## 9. Backfilling

Backfilling as specified under Section, EXCAVATING, FILLING, AND BACKFILLING, shall not be started before dampproofing has been completed and approved.

## 10. Payment

No separate payment will be made for the work covered under this section, and all costs in connection therewith shall be included in the lump-sum contract price for the structure to which the work pertains.

## NOTES TO CONTRACTING OFFICER

1. This specification is to be used in the preparation of contract specifications. It will not be made a part of a contract merely by reference; pertinent portions will be copied verbatim into the contract documents.

2. The section number will be inserted in the specification heading and prefixed to each paragraph and page number.

3. Paragraph 2: The listed designations for publications are those that were in effect when the specification was being prepared. These designations will be changed, as required, to those in effect on the date of invitation for bids; and the nomenclature, types, grades, classes, etc., referenced in the specifications will be checked for conformance to the latest revision or amendment. To minimize the possibility of error, the letter suffixes, amendments, and dates indicating specific issues will be omitted elsewhere in the specification. It is essential, therefore, that the list of applicable publications be retained in the contract specifications.

4. Paragraph 3: Where the drawings do not show all locations to receive dampproofing, this paragraph will be revised to include such additional locations. Interior dampproofing will be considered for other locations that will be subjected to frequent splashing of water in the normal use of the building.

5. Dampproofing of exterior walls will not be used for protection of subgrade work where a head of water or unusually wet soil conditions occur. In such locations membrane or metallic waterproofing will be specified and, unless applicable

to other portions of the work, the portions of this specification dealing with exterior damp-proofing will be deleted.

6. Paragraph 5: Sample requirements will be modified to meet contract specification requirements.

7. Paragraph 7: The hot-application type of exterior dampproofing as specified in paragraph 7a will be specified for all normal conditions and paragraph 7b deleted; but in exceptionally well drained soils the cold-application type of exterior dampproofing as specified in paragraph 7b may be specified, in which case paragraph 7a will be deleted. In either case, the in-applicable materials listed in paragraph 4 and the corresponding specifications in paragraph 2 will be deleted.

8. Concrete surfaces to receive dampproofing will be specified to have a smooth finish in Section CONCRETE, paragraph FINISHES. Masonry surfaces to receive dampproofing will be specified in Section MASONRY, to have joints cut flush.

9. Paragraph 10: The PAYMENT paragraph will be deleted from any specification contemplating one lump-sum contract price for the entire work covered by the invitation for bids.

# INDEX

AGO 10179A

www.ingramcontent.com/pod-product-compliance
Lightning Source LLC
Chambersburg PA
CBHW051216200326
41519CB00025B/7138

9 781410 108395